# PRODUCTION AND FEEDING OF SINGLE CELL PROTEIN

*Edited by*

M. P. FERRANTI

*Commission of the European Communities, Brussels, Belgium*

and

A. FIECHTER

*Institut für Biotechnologie,*
*ETH-Hönggerberg/HPT, Zurich, Switzerland*

APPLIED SCIENCE PUBLISHERS
LONDON and NEW YORK

APPLIED SCIENCE PUBLISHERS LTD
Ripple Road, Barking, Essex, England

*Sole Distributor in the USA and Canada*
ELSEVIER SCIENCE PUBLISHING CO., INC.
52 Vanderbilt Avenue, New York, NY 10017, USA

**British Library Cataloguing in Publication Data**

Production and feeding of single cell protein.
1. Proteins in human nutrition—Congresses
I. Ferranti, M. P.  II. Fiechter, A.
641.1'2  TP453.P7

ISBN 0-85334-243-1

WITH 71 TABLES AND 60 ILLUSTRATIONS



Organization of the workshop by the Commission of the European Communities, within the framework of the European Cooperation in the field of Scientific and Technical Research (COST) and the Federal Department of the Interior of Switzerland and the Institute of Biotechnology of the Swiss Federal Institute of Technology in Zurich.

Published for the Commission of the European Communities, Directorate-General Information Market and Innovation, Luxembourg.

EUR 8641



Printed in Great Britain by Galliard (Printers) Ltd, Great Yarmouth

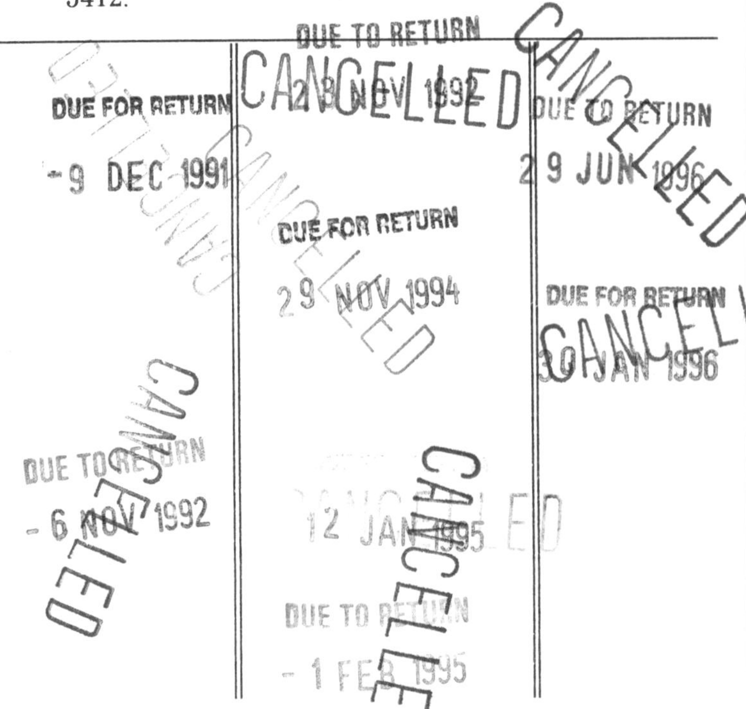

3

# PRODUCTION AND FEEDING OF SINGLE CELL PROTEIN

*Proceedings of the COST Workshop 83/84 on Production and Feeding of Single Cell Protein held in Zurich, Switzerland, 13–15 April 1983*

# FOREWORD

This workshop was organized within the framework of the European Cooperation in the field of Scientific and Technical Research (COST) and sponsored by the Commission of the European Communities, the Federal Department of the Interior of Switzerland and the Institute of Biotechnology of the Swiss Federal Institute of Technology in Zurich.

Since 1980, a research project on "Production and Feeding of Single Cell Protein (SCP)" (COST Project 83/84) has been implemented under a Memorandum of Understanding signed by Belgium, Denmark, Federal Republic of Germany, France, Ireland, The Netherlands, Spain, Sweden, Switzerland, Turkey and Yugoslavia. The three-year programme has led to a better understanding of the research potential in the participating countries and shown their ability to cooperate in this field.

Three workshops have been held during the project :

The project commenced with a workshop in Jülich from 15-16 October 1979, which indicated the state-of-the-art in the research areas to be investigated.

The workshop from 30 September to 2 October 1981 in Braunschweig brought together experts on chemistry and the structure of lignocellulosic materials and their modification by treatment and rumen microbiologists. An evaluation of promising new approaches was made and gaps in present knowledge were identified. The recommendations formed the basis for the definition of future work in a programme proposal for a follow-up of the COST-project.

The aim of the COST 83/84 workshop reported in this volume, held in Zurich from 13 to 15 April 1983, was to report on the activities of the past three years and to draw up evaluations and recommendations for future collaboration. Furthermore, guidelines for the Final Report of the COST 83/84 programme were settled.

The response to this workshop, in which over one hundred scientists from fifteen countries participated, and the number and quality of the papers showed the significance of the subject treated.

The workshop revealed that both technical and economic aspects have to be considered when strategies for the production of protein for animal feed from agricultural and forestry wastes are evaluated.

1. Lignocellulosic Materials
Progress has been made since the last workshop in Braunschweig in 1981, but the development of simple processes for farm application is not yet possible. More scientific knowledge will be necessary before less expensive and simpler pre-treatment and hydrolysis processes can be developed.

2. Whey
In contrast to the above mentioned topic, scientific background in the field of whey upgrading is available, but the economy of such a process depends on the conditions of each particular case. Only case studies will throw light on the economic aspects of such a process.

3. Nutritive Value and Toxicology
It has been shown that the nutritive value of lignocellulosic and whey SCP is excellent. Toxicological investigations do not give any evidence against the use of the produced protein as animal feed.

The conclusions of the Round Table discussion consisted of evaluations and recommendations to be submitted to the Commission of the European Communities, and in essence are as summarized above.

A proposal for a follow-up programme has already been submitted (COST 84 bis).

The full texts of the review papers and summaries of the short communications are collected within the present volume.

The organizers thank all the participants for having contributed their reports and for taking part in the discussions which led to the success of the workshop.

Brussels and Zurich
April 1983

M.P FERRANTI and A. FIECHTER

CONTENTS

SPECIAL SECTION - PRETREATMENT AND DEGRADATION OF LIGNOCELLULOSIC MATERIALS

## SUBJECT AREA 1 - PRODUCTION OF SCP ENRICHED SUBSTRATES FROM CELLULOSIC MATERIALS

## SPECIAL SECTION

## PRETREATMENT AND DEGRADATION OF LIGNOCELLULOSIC MATERIALS

Chairman : A. FIECHTER

2

# PRINCIPLES FOR PRE-TREATMENT OF CELLULOSE SUBSTANCES

By F. Rexen
Carlsberg Research Laboratory
Department of Biotechnology
Copenhagen. Denmark.

Summary

The aim of this paper is to give an up-to-date review of the nummerous pre-treatment methods for lignocellulosic biomasses that are described in literature. These methods may be placed in one or more of the following groups according to their main effect:
Removal of contaminants (lignin and hemicellulose): Pulping, oxidating agents biodelignification, steam treatment, pressure cooking, steam and freeze explosion.
Swelling of the cellulosic fibres: Alkali treatment.
Altering the crystallinity of the cellulose molecule: Grinding, selective dissolving of cellulose.
Depressing the degree of polymerization of the cellulose molecule: Acid treatment, extrusion.
None of the processes described are ideal for all conditions. The pre-treatment of choice will depend of the substrate selected, the end-product and the overall cost of pre-treatment.

## 1. INTRODUCTION

Plant biomasses besides cellulose contain lignin and hemicellulose in various amounts depending on the species. One of the functions of lignin and hemicellulose is to "protect" the cellulose against microbial attack and therefore all industrial processes aiming at microbiological and chemical conversion of cellulosic biomasses should include a pre-treatment step to render the cellulose more accessible.

The ideal pre-treatment would be cheap and have a low energy requirement, create no pollution problems and result in an increased cellulose content (i.e. removal of non-hydrolyzable materials), increased reactivity (i.e. decreased crystallinity) and increased bulk density.

A pre-treatment principle that meet all these demands has, however, not yet been found. Many pre-treatment methods have been proposed, but only a few will ever become commercially relevant, presumably as components in new systems. The task is highly complex. The pre-treatment procedure will depend on the quality of the raw material used (wood, straw, waste paper etc.) and the character of the end-product. For instance, if glucose is to be produced as an intermediate or final product, then a relatively pure cellulose substrate is required. On the other hand, crude single cell protein production and substrates used as energy feedstock need not to be purified or restricted to cellulose composition.

The availability and price of the raw material are evidently also decisive factors.

These factors depend not only on the actual production, but also on competing uses, collecting costs etc. For instance will there be many competing uses for wood, while collecting costs are rather high for straw. Rice hulls and bagasse are on the other hand produced in factories, and have therefore no collecting costs if the treatment plant could be combined locally. Also the alternative uses of these raw materials are limited (fuel and feed).

The special raising of crops for their cellulose content and biological availability is an aspect of the future. In Sweden the feasibility of energy farming has been studied. It is however doubtful whether this sort of agriculture will ever become economical, unless the energy crop at the same time produces high valued co-products. It should for example be possible to breed cereal crops that are optimized not only with respect to grain production, but also to cellulose content and quality especially in the straw fraction.

Today the choice of cellulosic raw material for the bioindustry is restricted to cellulosic wastes such as wood waste, straw and waste paper. These products contain cellulose, hemicellulose and lignin approximately in the proportion of 5:3:2. Hydrolyses of hemicellulose to mono- and oligo saccharides is comparatively easy, which can be accomplished with either acids or enzymes under moderate conditions.

Unlike hemicellulose, native cellulose is strongly resistent to hydrolysis. And this is roughly due to two reasons: It has a highly ordered crystalline structure, and a lignin seal surrounds the cellulose fibres as a physical barrier.

According to Tsao (34) the 1,4 β-glucosidic linkage in cellulose is from a process engineering viewpoint no more difficult to break than the 1,4 α-glucosidic linkage in starch if the cellulose molecules are fully hydrated and exposed and are free from hindrance by the lignin seal and the crystalline structure.

In other words, the difficulty in obtaining fast and complete hydrolysis of cellulose is not due to the primary linkage of the cellulosic chain, but rather the secondary and tertiary structure of cellulosic materials. An increase in the accessibility of cellulose may be obtained by a number of different processes, which all can be placed in one or more of the following groups according to their main effect: 1) Removal of contaminants (lignin and hemicellulose). 2) Swelling of the cellulose fibres. 3) Altering the crystallinity of the cellulose molecule. 4) Altering the degree of polymerization of the cellulose chain.

A final evaluation of pre-treatments should be based on costs of the pre-treatment process (preliminary grinding, steam, electricity, chemicals by-product recovery, waste disposal, capital costs and maintenance of equipment) and be related to yield of end-product.

## 2. REMOVAL OF LIGNIN AND HEMICELLULOSE

A separation of the cellulose fibers from lignin and hemicellulose is an efficient way of pre-treatment, which has been done commercially in the cellulose industry for many years. Cooked and bleached fibres with a low lignin content are comparatively easy to break down by microbes; the price of such fibres is however so high that it is prohibitive for use in a biotechnical process.

It is presumably not necessary to remove all the contaminating substances from the cellulose. It should be possible to estimate the optimal degree of separation from an economical point of view. Also for economic reasons is it important that the separated products, primarily the hemicelluloses are not completely destroyed in the separation process so that they can find use as chemical raw material, fuel or substrate for fermentation.

Alternative methods to delignification is the selective removal of hemicellulose followed by a second step to remove the lignin and/or cellulose.

2.1 Delignification

Apart from the chemicals used in the cellulose industry, many delignifying agents are known, and recent explorations have shown that lignin degrading organisms selectively may render the lignin fraction soluble (17, 8).

Lignin can also be selectively removed by treatment with oxidative chemicals such as $NaClO_2$ and ozone. Binder and co-workers (1) have investigated the effect of ozone on the biodegradability of straw. The ozone breaks down the lignin, and the dissolved lignin and other soluble components are separated from the solids by centrifugation.

It was demonstrated that the degree of delignification has a significant effect on the enzymatic degradability. Initially, however, is extensive delignification accompanied by only minor increases in digestibility. Following this lag phase digestibility rises rapidly with delignification, until it finally levels off at a value which depends on the enzyme activity.

A reduction of the lignin content beyond 50-60% of the original concentration does not enhance the microbiological degradability any further. The cellulose fraction was nearly quantitatively degradable by cellulolytic fungi, while the undigested straw could be digested only to a limited extent.

It was concluded that ozone is an efficient pre-treatment chemical; the cost of treatment is however so high that it is not feasible to use ozone, unless the price can be reduced or the efficiency of the process can be improved considerably.

Alternative oxidative degradation procedures of lignin have been studied. Moo Young et al. (12) investigated pre-treatment of aspen wood with sodium hypochlorite. The pre-treated product was fermented with the microorganism Chaetomium cellulolyticum, and a cellulose utilization as high as 90% was measured. The SCP produced contained 38% protein.

Millet et al. (21) have shown that also $SO_2$ is efficient in degrading lignin. They treated both softwood and hardwood with $SO_2$ under pressure and found that an essentially quantitatively enzymatic conversion of hardwood carbohydrates could be obtained, indicating a complete disruption of the lignin carbohydrate complex in the wood. The effect on softwood was less pronounced. Others have obtained similar results with sulphur dioxide (30).

It is useful to determine the relationship between the extent of delignification and enzymatic hydrolysis rate. As already mentioned the work by Binder et al (1) on ozone treatment showed that 50% delignification gave the highest yield of reducing sugar. This corresponds very well with findings by Fan et al. (10) who treated wheat straw with a variety of chemicals which yielded a wide spectrum of lignin content. They found that the hydrolysis rate increases substantially with the extent of delignification up to 50%; beyond this however, the hydrolysis rate increases only slightly. Similar observations have been reported by Millet et al. (21).

It seems that the effect of delignification depends on the rate of delignification, independently of how it is done, by oxidation, reduction etc. Fan and co-workers presume that delignification beyond 50% leads to a collapse in the lignocellulosic structure, thus shrinking the available surface area for enzymatic attack. Additionally, the cellulosic fibres might undergo structural rearrangement upon extensive delignification e.g. recrystallization of cellulose.

Work by Saddler et al. (31) showed that steam explosion treatment of aspen wood for 20 sec. followed by a mild oxidation with chlorite (2%) had a surprisingly strong effect on cellulose exposure. The steam explosion alone reduced the lignin content with approx. 40%, which is in accordance with other findings. The oxidation lowered the lignin content by only 1.6% but resulted in an additional 50% relative increase in reducing sugars on subsequent enzymatic hydrolysis.

The fact that removal of a small amount of lignin can have such an impact upon the cellulose accessability indicates according to Saddler that the lignin removed has been present as a relatively thin film of a large surface area.

A rather new aspect in biodelignification is the use of lignin degrading organisms as a pre-treatment. Detroy and coworkers (6) have f.ex. investigated the use of Pleurotus oestreatus for the pre-treatment of wheat straw.

The organism appeared to selectively degrade lignin during the first 6 days. By prolonged treatment, also the cellulose was degraded (70 days, 40% lignin and 32% cellulose are lost). Also Eriksons (9) experiments with lignin degrading fungi should be mentioned.

These lignin degrading organisms, that more or less selectively degrade the lignin fraction may lead to simplified pre-treatment technologies such as piles of wood chips or chopped straw being inoculated with the lignin degrading fungi or other microbial agents before the material is processed further.

### 2.2 Removal of hemicelluloses

As already mentioned, the hemicelluloses in biomasses are mostly polymers of pentoses (xylose and arabinose) and hydrolysis of the hemicelluloses to mono- and oligosaccharides is comparatively an easy matter. It can be accomplished by treatment with acids under mild conditions or with steam.

Detroy et al. (7) have reported that a pre-treatment of straw with steam at 170°C in 60 min. could remove up to 90% of the pentosans.

Puls and Dietrichs (25) describe a similar process which can remove almost completely the hemicellulose fraction in wood and straw. Besides a steam treatment at 170-200°C, the material was subjected to fiberizing by a flashlike expansion or in a refiner. The cellulose was not attacked but it was rendered more accessible to microbial degradation.

The so-called IOTEC process (38) is also a steam explosion process, which is similar to the old Masonite process for production of fiber boards. Biomass is heated at high pressure with steam to the point of structural softening. Then the pressure is released quickly, exploding the material and making the cellulose fibres more accessible to hydrolysis. Claims are made that this method can achieve 90% conversion of cellulose to glucose and 80% conversion of pentosans to xylose. The lignin isolated is said to be of high quality, suitable for making phenolic resins or other chemicals.

Another Canadian company, STAKE technology (33), has developed a continuous high pressure treatment process, which they call autohydrolysis, although there is no hydrolysis of the cellulose but rather a conversion of hemicellulose and lignin to more easily recoverable forms. In the STAKE digester, biomass is fed via a plug forming feeder into a steam pressurized cylinder containing a screw conveyor. The digester system is already in commercial operation at two locations, producing cattle feed.

In the above-mentioned treatments it is necessary to reach the temperature of softening (or melting) of the lignin, which is about 180°C. It seems that the modification of the lignin cellulose linkage is reversible and the treatment will be less effective if the treatment is prolonged. It is also important that the process conditions are not so severe that the by-products, mainly the hemicelluloses, are destroyed.

The steam explosion technique requires rather expensive equipment and thermal energy in the form of steam. It has the additional disadvantage that sugars are degraded by the high temperatures involved even at small reaction times.

Dale et al (6) have proposed an ammonia freeze explosion technique, called AFEX, which is said to overcome these problems. The technique relies on treatment with a low temperature boiling liquid under pressure followed by pressure release to evaporate the liquid and reduce the temperature. Liquid ammonia is preferred, as it also swells the fibres and to some extent decrystallizes cellulose, but also "inert" chemicals such as carbon dioxide may be used. In treatment of alfalfa stems and rice straw with ammonia (1 kg $NH_3$ per kg alfalfa) is achieved more than 90% conversion of cellulose to glucose by enzymatic hydrolysis. It is essential that most of the ammonia is recovered. It is claimed that only one percent remains with the fibre so that the recovery rate of 99% may be achieved.

## 3. SWELLING PRE-TREATMENTS

A number of chemicals may act as swelling agents. The chemicals are able to break the hydrogen bonds between the cellulose microfibrils, resulting in a considerable swelling of the material.

Aqueous solutions of relatively high concentrations of NaOH, ammonia, organic bases (amines) and certain salts ($ZnCl_2$ and $SnCl_4$) are chemicals of this type. The use of swelling agents have been quite successful in the area of upgrading the nutritive value of forages and wood residues for animal feed.

### 3.1 Sodium hydroxide

Sodium hydroxide has been used for pre-treatment of straw for feeding to ruminants since the beginning of this century, and the effect of sodium hydroxide treatment on rumen digestibility is well established.

Comparatively simple treatments such as soaking of straw in a dilute lye solution for 12 hours at ambient temperatures can increase the in vitro digestibility to 60-70% (Beckmann process).

Such treatments are however not applicable in industrial productions as they are water and time demanding and polluting - 20% of the straw dry matter is removed by washing. A so-called dry alkali treatment process, which is both cheap and efficient has been developed.

The process was developed in the beginning of the seventieth (27) and a number of plants have since then been built throughout Europe.

The product is used as a cattle feed, and the digestibility achieved is dependent on the amount of alkali used. Following correlation between organic matter digestibility and added amount of sodium hydroxide has been found: $OMD = 50.1 + 4.02\ x$ (28).

In the dry process, the straw is mixed with 4-6% NaOH in a concentrated solution, and afterwards pressed in a pelletpress under high pressure. The pellets are sold a comparably low price, and they constitute an interesting potential raw material for a coming biomass fermentation industry.

The pellets have been subject to a number of fermentation and hydrolysis experiments: Peitersen has for example (23, 24) in experiments with cellulase production from tricoderma Viride (QM 6a and QM 9123) used industrially produced straw pellets as substrate. He found that the cellulase yield was increased with 75% by using alkali treated pellets in stead of untreated straw. The fermentation time was reduced from 13 to 6 days.

During my time at Biotechnical Institute in Kolding (19) we tested industrially produced alkali treated straw pellets in fermentation with a mixed culture of a bacteria (Cellulomonas uda) and a yeast (Candida utilis). The straw pellets were milled (wet or dry) before fermentation, which was carried out both in continuously working fermentors and in batch fermentors. A SCP of 50-60% was achieved, and the fibres were decomposed to an extent of 60% at best.

We found that in average the industrially produced pellets are hydrolysed at a degree of 40% with cellulases (24 hours). The highest figure was 60%, corresponding to a conversion rate close the the theoretical figure.

### 3.2 Ammonia treatment

Ammonia is a weaker base than sodium hydroxide. Its ability to swell cellulosic fibres is however as effective as NaOH.

A number of farmscale ammonia treatment methods for improving the digestibility of straw used as cattle feed have been developed, and ammonia is widely used by farmers for this purpose. Fermentation experiments with ammonia treated biomasses have however shown that ammonia treatment is less efficient than f.ex. sodium hydroxyde treatment (14, 29).

Ammonia is therefore not likely to be widespread used as pre-treatment chemical for industrial fermentations, unless it is combined with pressure treatment, higher temperature etc.

## 4. ALTERING THE CRYSTALLINITY OF THE CELLULOSE MOLECULE

In average, native cellulose is 15% amorphous and 85% crystalline. The influence of crystallinity on the susceptibility of cellulose has been studied by many workers (22, 26, 37). Walseth (26) showed f.ex. that cellulolytic enzymes readily degrade the more accessible amorphous portions of regenerated cellulose, but are unable to attack the less accessible crystalline material. Treatments that would alter crystallinity include: reprecipitation from solution, mechanical disruption such as grinding or ionizing radiations.

It has been demonstrated that the crystalline residue left after enzymatic hydrolysis can be rendered again into an active form by physical treatment such as milling or heating (13).

### 4.1 Grinding

The idea of heat and milling treatments of cellulosic raw materials to reduce crystallinity was first introduced by Krupnova et al in 1963 (20). It has been observed by Ghose and Kostick (11) that hammer milling gives good size reduction and increased bulk density, but with no gain in susceptibility. Wet milling or beating is widely used in the pulp and paper industry, but the application of this technique as a pre-treatment for enzymatic hydrolysis is of very little effect. This was shown by Chang et al (3). Ball milling has until now shown to be best of the mechanical treatments. It reduces crystallinity and increases contact surface, and products with good availability and high bulk density may be achieved.

Fan et al (10) has compared the decrease in crystallinity after milling of wheat straw in 3 different types of mills. It is seen from figure 3 that ball milling (8 hours) is the most efficient both with regard to decrease in crystallinity and to enzymatic hydrolysis. Notice that the hydrolysis rate increased appreciably with a relatively small decrease in crystallinity.

Also continuous compression in roller mills has proven to increase cellulose accessability. Sugar yields of 48% were obtained from air dry newspaper after six passes through even speed pressure rollers (32). Roller mills are used in industries such as cane sugar factories, plastic industries, for metals rolling, ore chrushing etc.

4.2 Dissolving

A number of solvents can selectively extract cellulosics from lignocellulosic materials. It has f.ex. been known for years that an aqueous alkaline complex of iron and sodium tartrate can dissolve cellulose, and it has been used for analytical purposes for a long time. Also cadoxen - well known chemical complex - can dissolve cellulose. Cadoxen is an alkaline complex of Cadmium and ethylene diamine in water. When excess water is added to a cellulose-cadoxen solution, cellulose will precipitate as a soft flocculation. Upon standing cellulose can recrystallize, and it becomes again resistant to hydrolytic attack. When the cellulose is still in the form of a soft flocculation, it can easily be hydrolyzed with either acids or enzymes.

Fig. 4 gives a comparison of the rate and yield of hydrolysis of corn crop residue with and without pre-treatment with cadoxen (34).

Binder and Fiechter (1) have studied the effect of pre-treatment with an alkaline solution of sodium tartrate and ferric chloride on straw hydrolysis with cellulases, and Tsao and co-workers (35) have tested the effect on corn cobs. Binder and Fiechter found that with a weak enzyme preparation (FP activity: 0.3 IU), 60% of the cellulose could be hydrolyzed within 20 hours. This is a considerable increase compared to untreated material. A two step treatment procedure (treatment-hydrolysis-treatment-hydrolysis) turned out to be particularly effective. Over 90% of the cellulose in the straw was hydrolyzed within 4 hours.

The solvents are rather expensive chemicals, therefore is the recovery of the chemicals important for the overall process economy. Tsao has reported a recovery rate of 97% for sodium tartrate, which presumably is not enough to make the process feasible.

## 5. ALTERING THE DEGREE OF POLYMERIZATION OF THE CELLULOSE MOLECULE

The length of cellulose molecules in a fiber varies over a wide range from gamma cellulose containing less than 15 glucose units to alpha cellulose with as many as 10 000 glucose units per molecule. This variation could be expected to affect the rate of hydrolysis considerably, particularly by enzymes that digest the cellulose molecules from the terminal ends. Cowling (4) states however, that all strong acids and most isolated cellulases studied to date appear to hydrolyse cellulose at random along the length of molecules. In hydrolysis with acids DP drops quickly and levels off at a more or less constant value of 100-200. When the molecules become soluble, there will obviously be a great increase in susceptibility to enzymatic hydrolysis. But this change is more the result of solubilization than it is the direct result of chain shortening. Thus DP of itself probably is of limited significance in increasing the susceptibility of cellulose to hydrolysis. All pre-treatment principles that tend to shorten the cellulose chain are also hydrolysing the hemicellulose fraction.

Recently, dilute acid hydrolysis has been found to be an effective pre-treatment for newsprint, corn cobs, poplar and oak woods (17). After the pre-treatment, enzymatic hydrolysis gave glucose yields up to 100%. The pre-treatment consists of dilute acid hydrolysis carried out under mild conditions in a continuous plug flow reactor (1% sulphuric acid, temperatures less than 220°C, reaction time in the order of few seconds). Such pre-treatment conditions are milder than conditions used for complete acid hydrolysis of cellulose (at least 240°C, 1-2% $H_2SO_4$ or concentrated acid at ambient temperature.

Another recent approach to dilute acid hydrolysis has been brought to a reasonable, advanced state of development (38). Cellulosic material is fed continuously into a twin screw extruder of the type used for plastic fabrication. The extruder design allows the material to be compressed to about 30 atm. In the extruder's pressure section, minor amounts of sulphuric acid and steam are added. The acid catalysed hydrolysis requires only about 20 sec. About 50-60% of the cellulose is converted to glucose. The process is thus more a hydrolysis process than a pre-treatment process (38).

## 6. COMBINED PROCESSES

The ideal pre-treatment should as already mentioned include both an extraction of unwanted components and a decrease of crystallinity in the cellulose molecule. This ideal can not be met in one single pre-treatment step, but it is possible to approach this ideal by a two or more step pre-treatment.

Thus at Purdue University, USA, Tsao (35) has developed and tested a two step pre-treatment for sugar cane bagasse. In the first step, bagasse is treated with dilute sulphuric acid, whereby the hemicellulose fraction is solubilized into a crude mixture of fermentable sugars. Also most of the protein and mineral contents of the bagasse will be solubilized.

The liquid pentose stream is fermented with bacteria to produce 2.3 butanediol (0.5 kg per kg fermentable sugar).

The solid residue containing mostly lignin and cellulose is mixed with a cellulose solvent. A number of solvents have been tested and it was found that cadoxen iron/sodium tartrate and concentrated sulphuric acid were most efficient. For economic reasons, sulphuric acid, intimately mixed with the solid residue in a defiberizer, was preferred. The defiberizer provides a strong shearing action to break up the solids with the aid of the solvent. Immediately after fiberizing, the wet mixture is mixed with methanol to precipitate cellulose. The precipitate still contains lignin and cellulose, but the cellulose is by now in a highly disorganized amorphous form. The cellulose is again solubilized and hydrolysed by heating with additional water and separated from lignin by centrifugation. The hydrolysed product is finally fermented to ethanol. From 1 ton of bagasse is achieved 190 kg ethanol and 190 kg butanol. In fig. 5 is shown the effect of treatment on glucose yield.

## 7. FUTURE RAW MATERIALS FOR THE BIOTECHNOLOGICAL INDUSTRY IN EUROPE

Wood, waste paper and to some extent straw are today the most promising raw materials for a carbohydrate based bioindustry. We might however in near future reach a situation where also the whole cereal crop becomes an interesting possibility for the future biotechnical industry.

The production of cereal grains in EEC has within the last decade increased considerably (from 116 mill. tons in 1970/71 to 150 mill. tons in 1980/81), and we will presumably very soon become a net exporter of cereal grain, unless action is taken to increase the consumption within the Community. To export the surplus to countries outside EEC will be very expensive with the present EEC subvention policy, and it will increase the danger of food trade war between EEC and USA.

This situation calls for a new way of thinking. The traditional use of cereals for food and feed is not likely to increase any further within the Community for a number of reasons. Therefore, new ways of cereal utilization in industry should be found.

The individual fractions in a cereal plant have many theoretical applications, price and market relations have, however, limited the economical feasibility for most of these possibilities. The situation is changing and one might foresee a development that radically will change the price and competition relationships for a wide range of agriculture derived organic products. The fractions of a cereal plant may be used for production of a broad specter of commodities including fine chemicals, fermenting products, dietary products, sweeteners, thickeners, plastics, building materials, paper, etc.

One of the main issues for a future cereal derived technology should be to maximize the utilization of the whole cereal crop by an optimization of the application of each botanical fraction (straw, endosperm, germ, hull, etc.). Thereby the European grown cereal crop may become competitive with alternative imported raw material.

REFERENCES


1. BINDER, A. and FIECHTER, A. (1980). Solvent pretreatment of straw. COST Workshop on Production and Feeding of Single Cell Protein. Jü lich 1979.
2. CHAHAL, D.S., MOO YONG, M. and VLACH, D. (1981). Effect of physical and physiochemical pretreatments of wood for SCP production with Chaetomium Cellulolyticum. Biotechn. and Bioeng. Vol. 23: 2417-2420.
3. CHANG, M.M., CHOU, T.Y. and TSAO, G.T. (1981). Structure, pretreatment and hydrolysis of cellulose. Adv. in Biochem. Eng. 20. Springer Verlag Berlin: 16-40.
4. COWLING, E.B. (1975). Physical and chemical constraints in the hydrolysis of cellulose and lignocellulosic materials. Cellulose as a chemical and energy resource. Biotechn. Bioeng. Symp.: 163-182.
5. DALE, B.R. and MOEIRA, M.J. (1982). A freeze explosion technique for increasing cellulose hydrolysis. Biotechn. Bioeng. Symp. 12: 31-43.
6. DETROY, R.W., LINDENFELSER, L.A., JUHAIN, G.S. and ORTON, W.L. (1980). Biotechn. Bioeng. Symp. 10: 135-148.
7. DETROY, R.W., CUNNINGHAM, R.L., BOTHAST, R.J., BAGBY, M.O. and HERMAN, A. (1982). Bioconversion of straw cellulose/hemicellulose to ethanol. Biotechn. Bioeng. 24: 1105-1113.
8. ERIKSSON, K.E. (1974). Norsk Skogsindustri 5: 125.
9. ERIKSSON, K.E. (1980). Proceedings of the COST Workshop on Production and Feeding of Single Cell Protein. Jülich 1979: 48-63.
10. FAN, L.T., GHARPURAY and YONG HYUNG LEE (1981). Evaluation of pretreatments for enzymatic conversion of agricultural residues. Biotechn. Bioeng. Symp. 11: 29-45.
11. GHOSE, T.K. and KOSTICK, J.A. (1969). Symp. Adv. Chem. Ser. (ACS) 95: 415.
12. GHOSE, T.K. and KOSTICK, J.A. (1970). Biotechn. Bioeng. 12: 921.
13. GHOSE, T.K. (1977). Advances in Biochemical Eng. Springer Verlag, Ber-

lin: 39-76.
14. HAHN, Y.W. and CALLIHAN, C.D. (1974) Cellulose fermentation. Effect of substrate pretreatment on microbiological growth. Appl. Microbiology 27 No. 1: 159-165.
15. JURASEK, L. (1979). Developments in industrial Microbiology, 20: 177.
16. KING, K.W. (1963). Adv. in Enzymic Hydrolysis of Cellulose and related Materials. Pergamon Press, London: 159-170.
17. KIRK, T.K. (1975). Lignin degrading enzyme systems. Biotechn. Bioeng. Symp. 5: 139-150.
18. KNAPPERT, D., GRETHLEIN, H. and CONVERSE, A. (1980). Biotechn. Bioeng. 22: 1449.
19. KRISTENSEN, T.P. and REXEN, F.P. (1978). Eencelleprotein af halm. Medd. fra Bioteknisk Institut nr. 14.
20. KRUPNOVA, A.V. (1963). Lesokhim Prom. (USSR) 16: 8. Ref. by Ghose in Adv. in Biochemical Eng. Springer Verlag, Berlin.
21. MILLET, M.A. and BAKER, A.J. (1975). Pretreatment to enhance chemical, enzymatic and microbial attack of cellulosic materials. Biotechn. Bioeng. Symp. 5: 193-219.
22. NORKRANS, B. (1950). Physiol. Plant. 3: 75.
23. PEITERSEN, N. (1975). Production of cellulase and protein from barley straw by Trichoderma viride. Biotechn. Bioeng. 17: 361-374.
24. PEITERSEN, N. (1978). Fermentation of barley straw by Trichoderma viride. AIChE Symposium Series 74: 172.
25. PULS, J., DIETRICHS, H.H. (1980). OECD cellulose Program, Workshop 2: 125-133.
26. REESE, E.T., SEGAL, L. and TRIPP, V.M. (1957). Text. Res. Journ. 27: 626.
27. REXEN, F.P. (1972). Forøgelse af halms fordøjelighed ved kemisk behandling. Ugeskrift for Agronomer og Hortonomer 18: 364-365.
28. REXEN, F.P. and VESTERGAARD THOMSEN, K. (1976). The effect on digestibility of a new alkali treatment technique. Animal Feed Science and Technology 1: 73-83.
29. REXEN, F.P. (1977). Ammoniakbehandling af strå. Beretning nr. 78 fra Bioteknisk Institut.
30. REXEN, F.P. (1979). Fraktionering af celluloseholdige stoffer. Meddelelser fra Bioteknisk Institut 1: 22.
31. SADDLER, J.N., BROWNELL, H.H. and CLERMONT, L.P. (1982). Enzymatic hydrolysis of cellulose and various pretreated wood fractions. Biotechn. Bioeng. 24: 1389-1402.
32. TASSINARI, T.H., MACY, C.F. and SPANO, L.A. (1982). Technology advances for continuous compressing milling. Pretreatment of lignocollulosics for enzymatic hydrolysis. Biotechn. Bioeng. 24: 1495-1505.
33. TAYLOR, J.D. (1980). OECD cellulose Program, Workshop 2: 125-133.
34. TSAO, G.T. (1978). Cellulosic material as a renewable resource. Process Biochemistry, Oct.: 12-14.
35. TSAO, G.T. (1979). Pretreatment of cellulosic material for saccharification. Microbiology appl. to biotechnology. Proceedings 12. International Congress of Microbiology, Münich.
36. VERED, Y., MILSTEIN, O., FLOWES, H.M. and GRESSEL, J. (1981). Biodegradation of wheat straw lignocarbohydrate complexes. European Journ. of appl. Microbiology and Biotechnology, 12: 183-188 and 13: 117-127.
37. WALSETH, C.S. (1952). TAPPI. 35: 233.
38. WORTHY, W. (1981). Cellulose to ethanol projects loosing momentum. Chemical and Engineering News, 7, Dec.: 35-42.

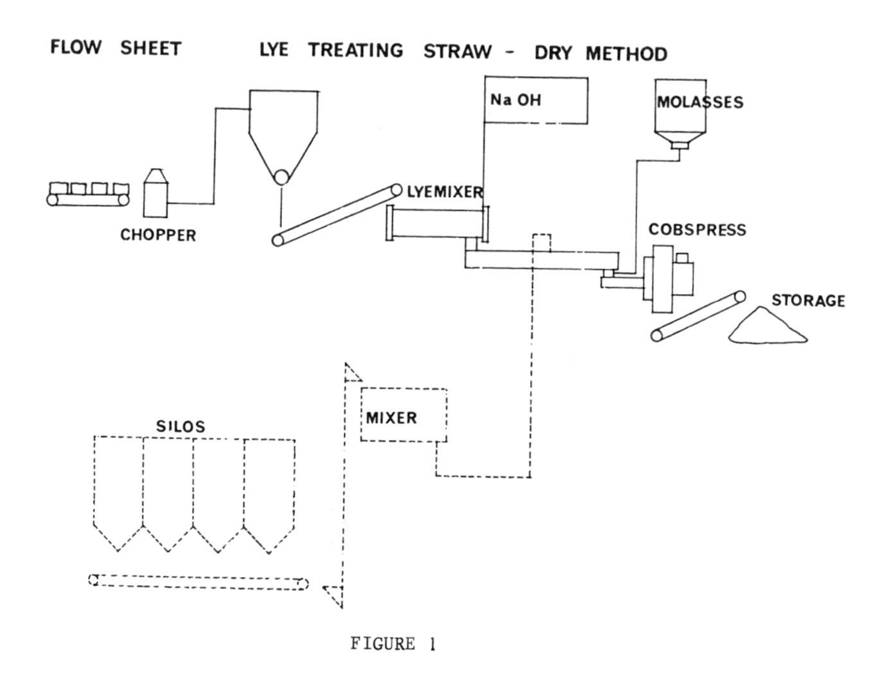


FIGURE 1

BATCH FERMENTATIONS IN A 500 L FERMENTOR WITH CELLULOMONAS SP. AND CANDIDA UTILIS (KRISTENSEN AND REXEN 1978)

| | PRETREATMENT - 6 %NAOH, PRESSING AND WET MILLING | | DRY MILLING - WASHING | |
|---|---|---|---|---|
| FERMENTATION TIME HOURS | 0 | 21 | 0 | 21 |
| TOTAL DRY MATTER % | 4,58 | 4,02 | 3,41 | 2,94 |
| FIBRE DRY MATTER % | 2,52 | 1,06 | 2,04 | 0,98 |
| CELL DRY MATTER % | - | 0,79 | - | 0,78 |
| CRUDE PROTEIN IN CELL DM % | - | 51,7 | - | 53,2 |
| FIBRE DECOMPOSITION % | - | 58 | - | 52 |
| CELL DRY MATTER G/G DECOMPOSED ORGANIC FIBERS | - | 0.55 | - | 0,76 |

FIGURE 2

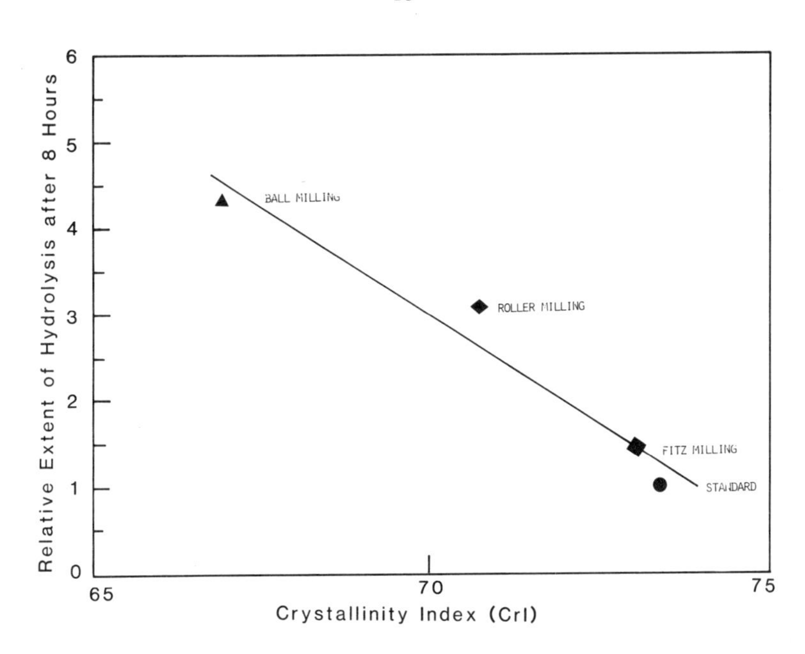


FIGURE 3

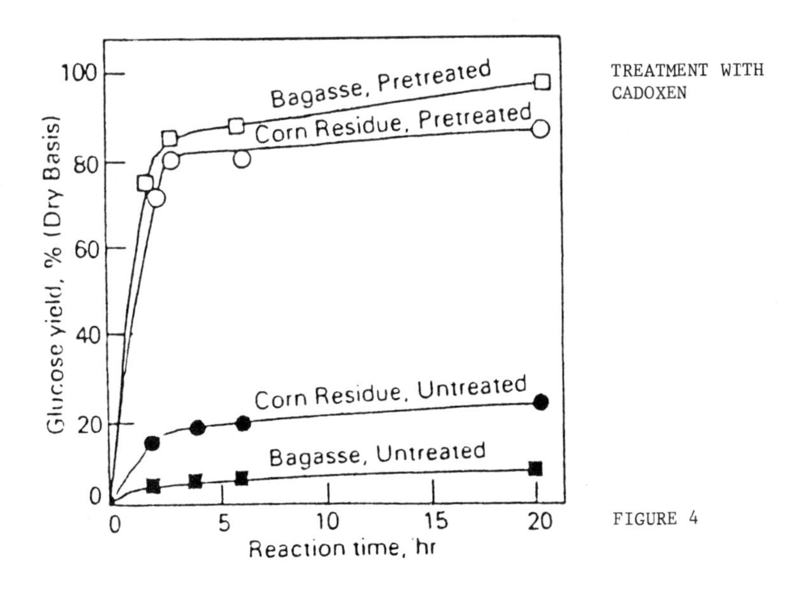


TREATMENT WITH CADOXEN

FIGURE 4

COMBINED PROCESS PURDUE

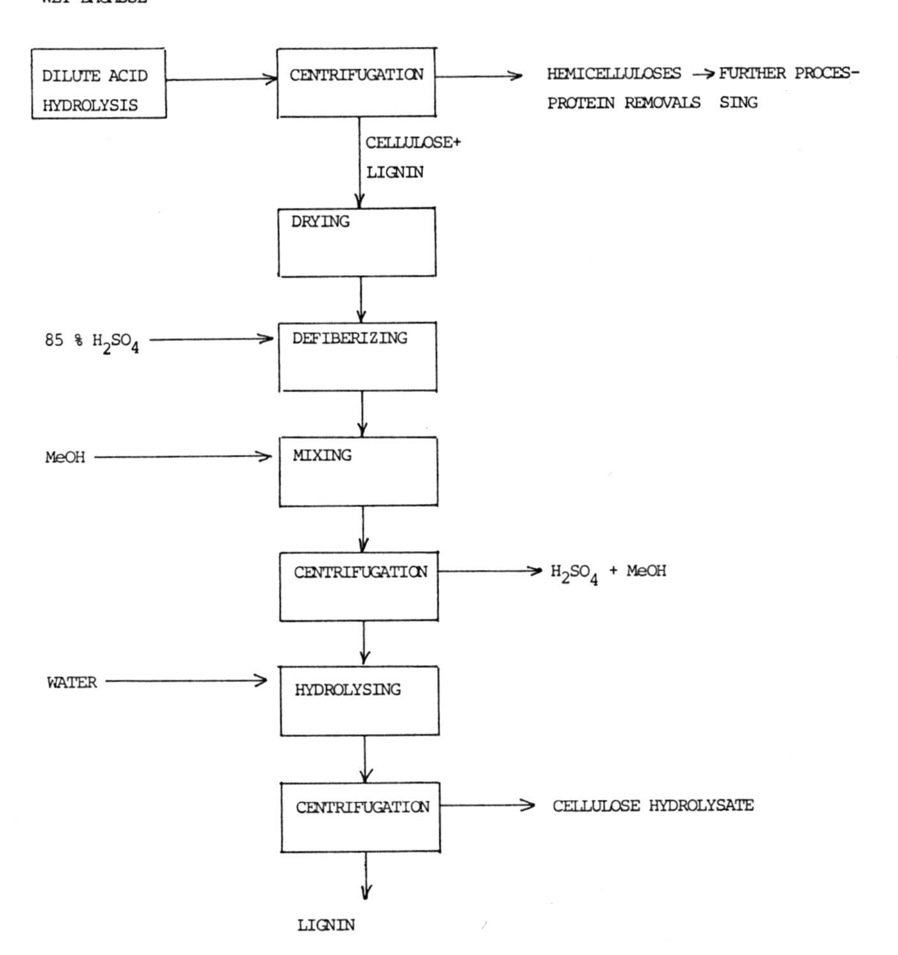


FIGURE 5

# 15

## PROSPECTS IN THE UNITED STATES FOR USING LIGNOCELLULOSIC MATERIALS

H. R. BUNGAY
Department of Chemical and Environmental Engineering
Rensselaer Polytechnic Institute
Troy, New York 12181, U.S.A.

Summary

The depressed economic situation in the United States and the drop in oil prices have led to lowered interest in fuels and chemicals from biomass. Nevertheless, research in the private sector and several years of support from governmental agencies have advanced several processes to the point where entrepreneurs are attempting to find capital to build factories. When all components of biomass lead to saleable products, process economics are attractive. The apportioning of costs to various products such as lignin leads to relatively inexpensive cellulose from which glucose can be obtained for making ethanol, single-cell protein (SCP), or other fermentation products.

## 1. INTRODUCTION

Biomass, the material produced by photosynthesis, is a potential feedstock for fuel and chemical industries. Various conversion technologies and their commerical prospects have been reviewed recently (1,2,3,4). The general status of biomass programs will be summarized, and several processes with good prospects for commercialization will be covered in more detail.

Liquid fuels for vehicles account for fifty per cent of the annual energy consumption in the U.S. The energy policy under the Carter Administration established a 1990 goal of 38 billion liters/year of ethanol for use in gasohol (10% ethanol, 90% gasoline). About forty per cent would have to come from non-grain feedstocks to avoid serious disturbances to food costs. Wood, municipal solid waste, crop residues, and perhaps crops grown especially for energy are alternate raw materials for producing alcohols, however, the technology for conversion was considered too long-range to make a contribution to the 1990 goal. Conversion of biomass to cheap fuels will not be reviewed here. The profit margins tend to be small and large-scale operations are essential when the value of the products is scarcely more than the cost of the feedstock. Better opportunities for profit exist in higher value products derived from the main components of biomass. There are few barriers to commercialization, and factories may be built in just a few years.

The composition of various woods and agricultural residues is:

| | |
|---|---|
| cellulose | 35 to 50 per cent |
| hemicellulose | 18 to 35 |
| lignin | 15 to 25 |

While cellulose and hemicellulose are polysaccharides that can be hydrolyzed to fermentable sugars, lignin has a three-dimensional phenolic structure that is highly resistant to microbial attack or to chemical hydrolysis. Indeed, it may be practical to use intact lignin in plastics, adhesives, and various formulations that make advantageous use of these properties.

A program managed by SERI (Solar Energy Research Institute, Golden, Colorado) was to accelerate development of biomass conversion to fuels or substitutes for chemicals now derived from petroleum. Research and development, process demonstration units (PDU's), prototype plant designs, marketing studies, and technology transfer were components of the plan. Other contracts for basic studies were to provide the foundations for more sophisticated advanced processes. Periodic reports about biomass projects funded by SERI are available (5).

Many of the process engineering contracts were terminated under the redirection of the FY81-82 alcohol fuels program. The philosophy now is that federally funded programs be aimed toward basic research and engineering categorized as high-risk and potentially high-payback, and that the private sector shall assume the risk and responsibility for commercializing the best technology from among competing processes. This point of view has merit because a number of projects have had sufficient success to attract private investment for further process development.

Although U.S. programs have focused on fuel alcohol, there may be more profitable uses for the sugars from biomass. Certainly, single-cell protein merits more attention, and an inexpensive source of glucose would result in quite attractive process economics. Sugars from hemicellulose are also accepted by some of the cultures employed for SCP.

## 2. RECENT PROJECTS

Most groups attempting to commercialize biomass processes need both scientists and engineers because fundamental and practical problems are intertwined. Genetic engineering of microorganisms holds promise for improving yields of enzymes that may also be subject to less feedback inhibition by products. High-rate processes may be possible using thermophilic organisms, and there are attempts to incorporate cellulase genes into bacteria and yeast to produce simultaneous hydrolysis and fermentation. Production of cellulase enzymes is presently very expensive, so higher yields of improved enzymes would impact strongly on profitability.

Better product recovery is another opportuntity for cost reduction. Distillation of ethanol to an anhydrous form can require energy equal to 25 to 50 per cent of the heat of combustion of the ethanol. Alternatives to distillation such as liquid/liquid extraction, membrane separation, and liquid and vapor phase adsorption could make a significant impact on both energy and cost for ethanol purification.

Recovery problems are minor for SCP because cells can be collected and dried by well established techniques. There is no need to invent new processes, and it should be easy to switch to pure glucose as the feedstock. The sugars from hemicellulose, however, could pose difficulties. The sugar intermediates obtained in biomass refining are contaminated by furfural, acetic acid, and a variety of organic compounds. However, byproduct sugar fractions from acid hydrolysis of wood are presently fermented to SCP in several countries.

Conversion processes based on cellulose hydrolysis may use acid or enzyme. Several groups have appreciated that acid hydrolysis may severely damage sugars formed early in the reaction, and sugars from hemicellulose suffer greatly during prolonged hydrolysis of cellulose. Sulfuric acid has been used because of its low cost, but the glucose yield is usually only about fifty-five per cent of the theoretical amount. Superior performance is obtained with other acids, but they are more costly and very efficient recovery is essential.

Although enzymes are very expensive and acid hydrolysis of cellulose may appear more economically attractive, enzymatic hydrolysis of cellulose has produced yields of over ninety per cent of theoretical, and prospects are excellent for the recovery of additional sugars from hemicellulose. As costs of enzymes continue to drop dramatically, enzymatic hydrolysis is likely to become highly profitable.

Hemicellulose can be removed easily by mild hydrolysis prior to a step for cellulose hydrolysis. Projects at Purdue University, University of California (Berkeley), Auburn University, M.I.T., and elsewhere are aimed at using the sugars from hemicellulose. These may be among the least expensive fermentation feedstocks. Hydrolysis conditions for the hemicellulose in wood chips are roughly 110 C with 0.1 per cent sulfuric acid. The unreacted solid residue could be dried and burned to supply energy.

Another approach to cellulose hydrolysis is to collect the sugars and recycle the unreacted solids. This would permit removal of the sugars from conditions that convert them to furfurals. However, all of the projects using acid hydrolysis seem to have ignored the commercial potential of lignin and have not considered whether it is damaged by the acid. Acid hydrolysis projects are summarized in Table I.

TABLE I: Acid Hydrolysis of Cellulose (5)

| Location | Comments |
|---|---|
| New York Univ. | Continuous twin-screw reactor with acid injection. Staged for close control. Requires rather fine particles. Yields in range of 55% of theory. |
| Dartmouth | Continuous slurry reactor. Very reliable determination of fundamental kinetics. |
| Auburn Univ. | Acid hydrolysis of hemicellulose and fermentation of resulting sugars. Good results on butanediol. |
| Purdue Univ. | Cellulose is softened with conc. sulfuric acid to destroy crystallinity. Hydrolysis yield is 95% with enzyme, 70% with acid but acid is cheaper. |
| Georgia Tech | Recycle of unreacted cellulose should shorten reaction with less degradation. |
| Berkeley | Two-stage nitric acid. Uses softwoods but can use hardwoods. High yields. |
| Michigan State Univ. | High yields with hydrogen fluoride. May require costly recovery and special materials of construction. |

## 3. BIOCONVERSION PROJECTS

The Solar Energy Research Institute has assembled teams to work on several aspects of processing of lignocellulosic materials. Mutational programs, screening cultures from around the world, and recombinant DNA techniques are used to develop superior cultures for producing enzymes or chemicals. From hot springs in Yellowstone Park, thermophilic anaerobes have been isolated that tolerate seven per cent ethanol; such an organism could lead to more rapid fermentations at elevated temperatures that permit easy recovery of ethanol by vacuum distillation directly from the fermentation broth.

The molds which produce cellulase have been studied at the U.S. Army Natick Laboratories for many years by Reese, Mandels, and co-workers, and these efforts plus contributions of other groups (especially at Rutgers University) have led to excellent strains that produce high titers of enzymes. A Berkeley process is derived from the Natick process and has investigated vacuum fermentation to improve productivity of the ethanol fermentation and has tried various approaches to better utilization of the components of biomass. The U.S. Army decided not to continue biomass energy projects, and the Natick team of scientists and engineers was disbanded and reassigned.

A process that originated at Gulf Oil Chemicals Company is being developed at the University of Arkansas. In order to eliminate feedback inhibition of the saccharification step, simultaneous fermentation of glucose to ethanol keeps the sugar concentration low. As with other enzymatic processes, a major cost is producing the cellulase enzymes.

There is no separate enzyme fermentation in a process developed at M.I.T. using carefully selected mixed cultures for direct hydrolysis of coarsely ground biomass. Two organisms are needed because one ferments glucose and is a good source of enzymes for hydrolyzing both cellulose and hemicellulose while the other is able to ferment hemicellulose monomers to ethanol. The process is so simple that unreacted residue costs little more than the raw material and can be burned to supply energy for the factory. Genetic improvements and strain selection were essential because the initial isolates had poor tolerance to ethanol and produced some lactic acid and about as much acetic acid as ethanol. New strains have higher ethanol tolerance and produce little acetic or lactic acids (6). Unfortunately, there is considerable variability in performance on different feedstocks (7). With corn stover, inhibitory products slow hydrolysis rates, impair enzyme stability, and acetate accumulates.

The Laboratory for Renewable Resource Engineering at Purdue University has greatly advanced understanding of the mechanisms of cellulose hydrolysis (8,9). A process has been proposed that makes use of concentrated sulfuric acid to soften cellulose and destroy its crystallinity. It is interesting that this facilitates both enzymatic and acidic hydrolysis; other pretreatments can have a large effect on one type of hydrolysis but little effect on the other. Products based on lignin have not yet been studied at Purdue, and it may be burned to power the factory if acid damages the lignin too badly.

Many of the ideas introduced for producing ethanol from cellulose would work equally well for SCP. For example, direct fermentation of cellulose by organisms with cellulase enzymes would eliminate the need for separate production of cellulase enzymes and the saccharification step. However, excellent conversion yields would be essential to avoid a mixture of unreacted cellulose and cells from which recovery of an acceptable product would be difficult.

## 4. PROCESSES THAT FOCUS ON LIGNIN

Removal of lignin from wood is well known, but several new processes have appeared that feature recovery of lignin as an important coproduct. The treatments are mild, so the lignin is reactive and has a relatively low molecular weight. Its properties are superior to those of lignin from the conventional Kraft or sulphite processes.

Collaboration among the University of Pennsylvania, Hahnemann Medical Center, Lehigh University, and the General Electric Company led to the formation of the Biological Energy Corporation. Their process is analogous to solvent pulping of wood chips, and extraction with acidic or alkaline alcohol dissolves lignin and hydrolyzes hemicellulose to soluble sugars. The cellulose that remains has long fibers and should find use in the highest grades of paper. During alcohol recovery, the lignin precipitates and is recovered for chemical uses, probably in plastics such as phenol-formaldehyde. The aqueous residue is a solution of mixed sugars rich in pentoses that are suitable for feeding cattle. Another possible product is the leafy residue if trees are harvested while green. Protein content is about 23 per cent., and similar material is sold in Europe as cattle feed. Fertilizer would probably be required to maintain soil fertility if the entire tree is harvested.

The Battelle/Geneva process is quite similar to the scheme of the Biological Energy Corporation except that the solvent for lignin is phenol (10). Phenol is miscible with boiling water and the low pH hydrolyzes hemicellulose as lignin dissolves. The solid residue is enriched cellulose with fibrous structure.

The aqueous phenol extract separates into two phases when cooled. The phenol-rich phase has the lignin, and the water-rich phase has the sugars from hydrolysis of the hemicellulose. Phenol is recovered and recycled. Phenol to replace process losses could be made by hydrocracking of lignin. However, preliminary discussions with companies interested in hydrocracking have indicated that other applications may command a higher price for lignin. Most uses for the sugars from hemicellulose could require very efficient removal of residual phenol.

Another solvent for lignin has been reported (11). An aqueous solution of eighty per cent formic acid at 110 to 120 degrees extracts most of the lignin in less than one hour. It is claimed that solvent recovery is easy and that wood can be processed as chips and does not have to be ground. However, formic acid is quite corrosive so the reactor must be constructed of costly materials. The cellulose should have excellent fiber characteristics that command a premium price for paper. Again, lignin would precipitate during solvent recovery.

All of these solvent extraction processes have a fibrous cellulose that should be highly desirable for paper, and there is no incentive for hydrolyzing it to glucose. The main fermentable fraction from such biomass refining is derived from hydrolysis of hemicellulose.

The recession has dampened enthusiasm for new ventures in the United States. Refining of biomass with solvents seems highly attractive, but interest may fade unless some large paper companies provide financial support. It may be a case of the main idea being around for a long time, and there may not be sufficient appreciation that innovations and the economic potential of lignin products have created a fresh situation and new opportunities.

The outlook is different for steam explosion processes for disintegrating biomass because the cellulose is damaged so that its main value is for hydrolysis to glucose. Steam-explosion in a continuous mechanical screw device has been commercialized for converting wood to cattle feed by Stake Technology of Canada. It should be possible to refine this material to get lignin and fermentable sugars.

An all-aqueous process for refining biomass has been developed by the Iotech Corporation of Ottawa. Commercial wood chips are impregnated with high pressure steam, and sudden release of pressure disintegrates the structure and potentiates enzymatic hydrolysis of the cellulose. About seven per cent of the exploded wood goes to the production of cellulase enzymes. Cellulase broth is added directly to the rest of the wood to achieve better than 85 per cent of the theoretical glucose from cellulose. Fermentation products such as ethanol or SCP are future targets, but capital costs will be lower and plant start up will be easier when the initial carbohydrate product is syrup. Solid residue from enzymatic hydrolysis is rich in lignin that can be recovered in quite pure form by extraction with dilute alkali, or the residue can be sold directly for adhesive applications such as binding of plywood or chipboard where some inert fillers are desirable in the lignin.

Tests of the Iotech process at a 1000-liter scale at the Gulf Oil Chemicals pilot plant in Kansas achieved about ninety per cent of the theoretical glucose from cellulose, and fermentation to ethanol performed well (5). Recipes for wood binders using the lignin have received successful evaluations by potential customers. There may be projects funded by New York State leading to a decision about locating a plant there.

All processes evolve, and some examples taken from Iotech may be of interest. The main capital and operating costs are for production of cellulase enzymes, thus much research and development are aimed at improving yields and optimizing the subsequent step for saccharification to minimize enzyme dosage. The enzyme fermentation is expensive because the organism Trichoderma reesei grows slowly, uses large amounts of oxygen, and elaborates an enzyme that must be used in high dosages because it turns over more slowly than a comparable commercial enzyme, amylase. To insure asepsis throughout this long fermentation, careful techniques must be employed, and the equipment design must be rigorous with respect to minimizing sources of contamination.

The initial cost estimates for the saccharification step were based on clean but not sterile operations. However, it was found at the Kansas pilot plant that several saccharification runs were highly successful and then serious trouble was encountered. The rich nutrition provided by copious amounts of sugars plus some minerals and nitrogenous ingredients greatly favored microbial growth. Once foreign organisms became established, they were almost impossible to control. A few good runs were performed with antibiotics, but soon resistant strains developed. From that point on, the saccharification runs were mostly disasters. Nevertheless, the pilot plant does not mimic a true production operation because materials were allowed to stand around. Following steam explosion, the disintegrated wood is sterile. It should be possible to conduct the washing and extracting steps with clean techniques and to initiate hydrolysis with very few organisms present. If it is possible to avoid aseptic techniques and sterilized pressure vessels, the savings in construction costs would be significant.

Just as the LORRE group (Laboratory of Renewable Resource Engineering) at Purdue is willing to sacrifice yields by using acid hydrolysis instead of expensive enzymatic hydrolysis, the Iotech group could build their initial factory minus the enzymatic steps. The abbreviated Iotech process is shown in Figure 1. By changing to a modified steam explosion with catalyst, all of the hemicellulose and part of the cellulose are hydrolyzed. The resulting sugars can be washed from the solid residue with water, and then evaporation would give a syrup to be sold. This is analagous to the concept mentioned previously of extracting the easy hemicellulose sugars, but the Iotech conditions derive some glucose as well, and the lignin extracted from the residue has been shown to have excellent properties for use in wood adhesives. While not as good in terms of sugar yield as some of the acid hydrolysis processes now under development, this Iotech scheme is better than existing methods as shown in Table II.

This embodiment of the Iotech process has lignin as the main product. Residual cellulose has little value because its degree of polymerization is too low for chemical applications such as making cellulose acetate. However, the fact that it can be so easily hydrolyzed in high yields with

TABLE II: Comparison of Hydrolysis Yields

| | % of Theoretical | % of Feedstock |
|---|---|---|
| Sulfuric Acid Hydrolysis | 55 | 25 |
| Dilute Acid for Hemicellulose | 90 | 18 |
| Iotech Wash of Wood | 80 (Hemicellulose)<br>30 (Cellulose) | 32 |

enzymes would suggest that this cellulose should be an excellent substrate for the production of SCP by an organism with cellulase activity. The conversion yield may be so high that the unreacted cellulose may not have to be removed from the protein.

Eventually, the Iotech process will probably employ enzymatic hydrolysis, and recycle of the enzyme will save on its production cost and permit higher concentrations for saccharification (12). Recycle is based on the affinity of cellulases for cellulose and on ways of reducing this affinity. Of the dozen or so components of cellulase, several bind strongly to cellulose. These can be recovered easily from the hydrolysate by adsorption on cellulosic material that can be blended with the solids fed to hydrolysis. Enzymes on the solid residue from hydrolysis can be desorbed with dilute buffer (13), and the extract would supplement the fresh enzyme. Already there are indications that about fifty per cent of the cellulase can be recycled, and the impact on costs will be dramatic.

## 5. CONCLUSION

Although support from the Federal government has been redirected toward long-range topics, a sound foundation has been laid for commercialization of biomass conversion. The first factories are likely to emphasize high value products such as lignin, molasses, and paper pulp, but the establishment of refining technologies will provide inexpensive fermentable sugars. A major factor for SCP could be a cellulose residue after the hemicellulose and lignin are gone. The residue from the Iotech process can be fairly pure cellulose of a low degree of polymerization that should be fermented easily by organisms that produce cellulases. If fermented directly, a simple, highly profitable SCP process should result.

## REFERENCES

1. BLANCH, H.W. and WILKE, C.R. (1982). Sugars and chemicals from cellulose. Reviews in Chem. Engr. 1: 71-119.

2. BUNGAY, H.R. (1982). Biomass Refining, Science, 218: 643-6.

3. BUNGAY, H.R. (1983). Commercializing biomass conversion. Environ. Sci. Technol., 17: 24A-31A.

4. NG, T.K., BUSCHE, R.M., McDONALD and HARDY, R.W.F. (1983). Production of feedstock chemicals. Science, 219: 733-40.

5. Alcohol Fuels Program Technical Review, available by contacting Dr. Larry Douglas, SERI, 1617 Cole Blvd., Golden, CO 80401.

6. AVGERINOS, G.C. and WANG, D.I.C. (1980). Direct microbial conversion of cellulosics to ethanol. Ann Rep. Ferm. Proc. 4: 165-91.

7. JOHNSON, E.A., SAKAJOH, M., HALLIWELL, G., MADIA, A. and DEMAIN, A.L. (1982). Saccharification of complex cellulosic substrates by the cellulase system from Clostridium thermocellum. Appl. Environ. Microbiol. 43: 1125-32.

8. CHANG, M.M., CHOU, T.Y.C. and TSAO, G.T. (1981). Structure, pretreatment, and hydrolysis of cellulose. Adv. Biochem. Engr. 20: 15-42.

9. GILBERT, I.G. and TSAO, G.T. (1983). Interaction between solid cellulose substrate and cellulase enzymes in cellulose hydrolysis. Ann. Rept. Ferm. Proc. 6: (in press).

10. KRIEGER, J. (1982). Process yields marketable biomass fractions. Chem. Engr. News, May 31, 1982.

11. BUCHOLTZ, M. and JORDAN, R. (1981). Formic acid wood pulping: chemicals and fuels. Paper presented at 3rd Annual Energy Seminar, Gannon Univ., Erie, PA.

12. SINITSYN, A.P., BUNGAY, H.R. and CLESCERI, L.S. (1983). Enzyme management in the Iotech Process. Biotechnol. Bioengr. (in press).

13. SINITSYN, A.P., BUNGAY, M.L., CLESCERI, L.S. and BUNGAY, H.R. (1983). Recovery of enzymes from the insoluble residue of hydrolyzed wood. Appl. Biochem. Biotechnol. 8: 25-29.


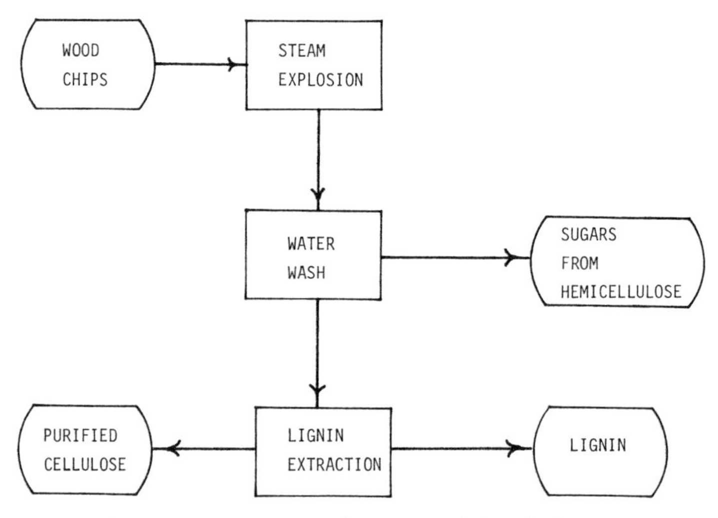


Fig. 1 - Simplified Iotech Process (No Saccharification)

# BIODEGRADATION OF LIGNIN BY PHANEROCHAETE CHRYSOSPORIUM

M.S.A. LEISOLA
Lignocellulose Group, Department of Biotechnology
Swiss Federal Institute of Technology, ETH-Hönggerberg
CH-8093 Zürich

## 1. INTRODUCTION

The white-rot fungi that belong to Basidiomycetes are probably the most efficient of all known lignin degraders. A considerable amount of information is now available on the physiological conditions and biological mechanisms of lignin degradation. Most of this information has been obtained from studies done with the white-rot fungus Phanerochaete chrysosporium (Sporotrichum pulverulentum). This report summarizes the work done with this fungus at the Department of Biotechnology, ETH-Zürich, with reference to the recent literature.

## 2. LIGNIN CAN BE RAPIDLY DEGRADED

Phanerochaete chrysosporium degrades lignin efficiently only under nonagitated conditions. However, these conditions lead to poor oxygen transfer into the mycelial mat (1, 2). By avoiding partially oxygen-starved conditions, we have been able to increase the rates of lignin degradation by this organism. Using various isolated lignin preparations, as much as 1 g lignin per 1 g cells can be degraded in a day (3) during the active phase. To our knowledge, this is at least 5 times more than the earlier reported results (4). It seems that the major obstacle for rapid biological degradation of lignocellulosic materials is not so much the structure of lignin, but the structure of wood itself.

## 3. GROWTH PATTERN OF PHANEROCHAETE CHRYSOSPORIUM

This organism degrades lignin only in nitrogen-starved conditions (5). Under these conditions, mycelial and cell-bound polysaccharides are also produced. The organism is able to recycle its own nitrogen, which is common for wood-rotting Basidiomycetes (6) and establish new growth at 10-15 day intervals, even under nitrogen-starved conditions. Lignin is not degraded during the growth phase but only during the rest phase (7).

Under laboratory conditions with excess glucose, the dry weight of the cultures increases periodically, while the old cell wall polysaccharides are not degraded. Thus, although the cell mat remains active for at least up to 2 months, the transfer of $O_2$ into the culture becomes increasingly more difficult with the ever-thickening cell mat.

## 4. EFFECT OF ATMOSPHERE ON LIGNIN DEGRADATION

The ability of P. chrysosporium to recycle its own nitrogen and establish new growth under nitrogen-starved conditions shows that the organism is able to survive under extreme conditions, which it also faces on and in wood materials in nature. According to our experiences, P. chrysosporium is able to efficiently degrade lignin under atmospheres of up to 30% $CO_2$ and down to 10% $O_2$. This is contrary to many reports in the literature that claim a 60-100% oxygen atmosphere enhances lignin degradation (8, 9). We believe that the major reason for this interpretation is the difficulty of oxygen transfer into the mycelial mat, rather than the composition of the atmosphere.

## 5. MECHANISM OF LIGNIN DEGRADATION

There is increasing evidence that the breakdown of lignin polymers occurs fairly unspecifically through, e.g., hydroxyl radicals (·OH) (10, 11). However, the system shows some specificity as only aromatic rings with three hydroxyls can be split by the organism (12). Many different types of lignins and lignin-related polymers can be degraded by the ligninolytic system of the fungus (3).

## 6. PRODUCTS OF LIGNIN DEGRADATION

The major product of lignin degradation is $CO_2$. However, there are many reports of low molecular-weight degradation products resulting from fungal attack on wood (13, 14), isolated lignins (3), and synthetic lignins (15). In contrast to lignin, many of these compounds are water-soluble and have an acidic nature and an average molecular weight of about 1000. We have noticed that similar products are produced from many different types of lignins (3). The proportion of $CO_2$ to soluble products can be affected, for instance, by energy conditions. More water-soluble material is produced when the culture is oxygen-limited or glucose as a co-substrate is replaced with cellulose or wood. Chen and co-workers (13) have studied the soluble products and found a large amount of different compounds. The open question is whether the process can be controlled to produce only certain types of molecules or whether it is totally non-specific and will always result in the production of a very large number of different types of molecules.

## REFERENCES


1. LEISOLA, M., ULMER, D. and FIECHTER, A. (1983). Problem of oxygen transfer during degradation of lignin by Phanerochaete chrysosporium. Eur. J. Appl. Microbiol. Biotechnol. (in press).

2. LEISOLA, M., ULMER, D., HALTMEIER, T. and FIECHTER, A. (1983). Rapid solubilization and depolymerization of purified Kraft lignin by thin layers of Phanerochaete chrysosporium. Eur. J. Appl. Microbiol. Biotechnol. (in press).

3. ULMER, D., LEISOLA, M., SCHMIDT, B. and FIECHTER, A. (1983). Rapid biodegradation of various isolated lignins by Phanerochaete chrysosporium. Appl. Environ. Microbiol. (in press).

4. JEFFRIES, T.W., CHOI, S. and KIRK, T.K. (1981). Nutritional regulation of lignin degradation by Phanerochaete chrysosporium. Arch. Microbiol. 117:277-285.

5. FENN, P. and KIRK, T.K. (1981). Relationship of nitrogen to the onset and suppression of ligninolytic activity and secondary metabolism in Phanerochaete chrysosporium. Arch. Microbiol. 130:59-65.

6. MERRIL, W. and COWLING, E.B. (1966) Role of nitrogen in wood deterioration: Amounts and distribution of nitrogen in free stems. Can. J. Bot. 44:1555-1580.

7. ULMER, D., LEISOLA, M., PUHAKKA, J. and FIECHTER, A. (1983). Phanerochaete chrysosporium: Growth pattern and lignin degradation. (to be published).

8. BAR-LEV, S.S. and KIRK, T.K. (1981). Effects of molecular oxygen on lignin degradation by Phanerochaete chrysosporium. Biochem. Biophys. Res. Comm. 99:373-378.

9. REID, I.D., SEIFERT, K.A. (1982). Effect of an atmosphere of oxygen on growth, respiration, and lignin degradation by white-rot fungi. Can. J. Bot. 60:252-260.

10. FORNEY, L.J., REDDY, C.A., TIEN, M. and AUST, S.D. (1982). The involvement of hydrogen peroxide in lignin degradation by the white rot fungus Phanerochaete chrysosporium. Arch. Microbiol. 130:59-65.

11. KUTSUKI, H. and GOLD, M.H. (1982). Generation of hydroxyl radical and its involvement in lignin degradation by Phanerochaete chrysosporium. Biochem. Biophys. Res. Comm. 109:320-327.

12. BUSWELL, J.A. and ERIKSSON, K.E. (1979). Aromatic ring cleavage by the white-rot fungus Sporotrichum pulverulentum. FEBS Lett. 104: 258.

13. CHEN, C.L., CHANG, H.M. and KIRK, T.K. (1982). Aromatic acids produced during degradation of lignin in spruce wood by Phanerochaete chrysosporium. Holzforschung 36:3-9.

14. REID, I.D., ABRAMS, G.D. and PEPPER, J.M. (1982). Water-soluble products from the degradation of aspen lignin by Phanerochaete chrysosporium. Can. J. Bot. 60:2357-2364.

15. CHUA, M.G.S., CHOI, S. and KIRK, T.K. (1983). Mycelium binding and depolymerization of synthetic $^{14}C$-labelled lignin during decomposition by Phanerochaete chrysosporium. Holzforschung (in press).

# PROGRESS AND PROBLEMS IN THE UTILIZATION OF CELLULOSIC MATERIALS

M. LINKO
VTT Biotechnical Laboratory
Tietotie 2 Espoo 15 Finland

Summary

The problems seen in the enzymatic hydrolysis of cellulosic materials have not changed much during the last five years. However, there has been important progress towards solving many of these problems. One of the key factors in terms of feasibility is the enzyme cost. The development of enzyme production processes has been quite impressive.

Five years ago I presented the following list of problems encountered in the enzymatic hydrolysis of cellulosic materials (1):

- Expensive raw materials (incl. collecting, transport etc.),
- raw material resistant to enzymatic hydrolysis,
- expensive pretreatment to improve reactivity,
- not all major components of the raw material efficiently utilized (cellulose, hemicellulose, lignin),
- expensive enzyme preparations,
- much enzyme needed,
- limited possibilities for enzyme reuse,
- slow hydrolysis,
- incomplete hydrolysis,
- enzyme preparation not optimally balanced (lack of some component, e.g. β-glucosidase),
- dilute substrate slurry necessary for efficient hydrolysis,
- inhibiting compounds present (e.g. in some industrial wastes),
- cheap final products (e.g. ethanol, SCP).

This list is still pertinent and many of these problems are severe. However, there has also been quite significant progress during the last five years.

The basis for development of processes for utilization of cellulosics is the vast, ubiquitous renewable raw material source. Therefore, it may sound paradoxical that the cost and even availability of raw material is a major problem, but it really is, for a number of reasons such as alternative uses, collection and transport. There has been some progress in the development of fast-growing tree cultivations, especially in some tropical or subtropical countries. Promising results have been achieved e.g. in the Philippines with several species of trees such as Leucaena leucocephala, Gmelina arborea and Eucalyptus deglupta (2). In this connection it should be pointed out that the areas used for tree plantations should not be high quality arable land.

The next points on the list of problems are still in force: the cellulosic materials are resistant to hydrolysis and efficient pretreatments tend to be costly. Furthermore, the high silica content of some materials such as sugar cane bagasse or rice hulls causes additional handling problems. One sensible approach is to admit the difficulties in complete hydrolysis of cellulosic materials and to be content with breaking down the easy part only, i.e. primarily hemicellulose and the amorphous regions of cellulose. An example of advantageous incomplete hydrolysis is the treatment of silage with cellulolytic enzymes (3).

New potential is seen in the utilization of hemicellulose, not only for SCP but also for ethanol production. There has been important progress in the ethanol fermentation of pentoses. The mold Fusarium oxysporum is one of the most promising organisms (4). Many microorganisms produce xylanolytic enzymes. Trichoderma reesei, well known for its production of cellulases (5), is also an efficient producer of xylanase.

It has been suggested that the xylan fraction should be first hydrolyzed with acid in relatively mild conditions, followed by enzymatic hydrolysis of cellulose (6). Simultaneous acid hydrolysis of xylan and cellulose is disadvantageous, because xylose is largely decomposed in the severe conditions necessary for hydrolysis of cellulose.

Progress in the production of cellulases and xylanases has been remarkable during the last five years. The famous Natick strain, Trichoderma reesei QM 9414, is a very stable mutant which produces cellulases and other enzymes rather efficiently (5), but this strain has lately been superseded by new ones capable of producing considerably higher enzyme activities (7,8,9).

Besides Trichoderma reesei some other fungi and also bacteria hold promise as enzyme producers. For example, an advantage of a cellulolytic enzyme preparation from Cellulomonas sp. is its resistance to inhibition by xylose, glucose, cellobiose and ethanol (10). This bacterium also produces xylanase and β-xylosidase (11). The advantage of Thermomonospora sp., a filamentous bacterium, is the thermostability of the cellulases produced. An efficient new mutant of this bacterium has recently been developed (12).

Production of β-glucosidase, either simultaneously or in a separate process is no longer a problem. The β-glucosidase of Aspergillus niger is less sensitive to end product inhibition than that of Trichoderma reesei (13).

Advances in research on the biochemistry of cellulases and hemicellulases may stimulate progress in the enzyme production and hydrolysis technology. The advancement of enzyme production technology includes alternative carbon sources such as lactose (9,14) or cheap pretreated and washed cellulosic materials (15), temperature and pH gradients during the production process (16), fed batch (17,18,19), semi-continuous (20) or continuous fermentations (16) and immobilized cells (21). All these advances have led to productivities exceeding 100 IU $l^{-1}h^{-1}$, whereas typical satisfactory productivities five years ago were about 30 IU $l^{-1}h^{-1}$. Consequently, the enzymes can be produced at a significantly lower price. However, in spite of this progress the problem of high enzyme cost still remains. An essential drawback is the low specific activity of cellulases, compared e.g. with amylases. Much enzyme is needed and the reuse of enzymes is difficult.

The enzymatic hydrolysis is slow and always will be slow compared with acid hydrolysis. On the other hand, enzymatic hydrolysis can be more complete than the acid hydrolysis, in which a considerable amount of sugars are decomposed, often to inhibitory compounds such as furfural.

If ethanol is the final product, stimulation of the enzymatic hydrolysis can be achieved by combined hydrolysis and fermentation. Either the conventional yeast Saccharomyces cerevisiae (22) or the bacterium Zymomonas mobilis (23) can be used for ethanol fermentation. The combined hydrolysis and fermentation also decreases the danger of microbial contaminations. On the other hand, the environmental conditions must be a compromise. For example, the temperature must be lower than in a separate hydrolysis process.

Dilute substrate slurries are hard to avoid in a conventional hydrolysis procedure. The situation is different in the treatment of silage with cellulolytic enzymes.

Both ethanol and SCP are cheap final products and will probably remain relatively cheap. For production of animal feed, the treatment of silage with cellulases, or some kind of solid state fermentation may be more feasible alternatives than a relatively complete hydrolysis of cellulosic materials to sugars, combined with or followed by SCP production.

## REFERENCES


1. LINKO, M. (1978). An evaluation of enzymatic hydrolysis of cellulosic materials, Proc. XII Intern. Congr. Microbiol., München, Dechema-Monographien Vol. 83, 209-218.
2. KETOLA, P. (1983). Manila Seedling Bank Foundation, Inc., Quezon City, Philippines. Personal communication.
3. VAISTO, T. et al. (1978). The use of cellulases for increasing the sugar content of AIV silage. J. Sci. Agric. Soc. Finland 50, 392-397.
4. ENARI, T-M. and SUIHKO, M-L. (1982). Ethanol production by fermentation of pentoses and hexoses from cellulosic materials. Symp. Fundamental and Applied Aspects of Ethanol Production by Yeasts and Other Microorganisms. Boston. CRC Critical Reviews in Biotechnology, in press.
5. MANDELS, M. et al. (1971). Enhanced cellulase production by a mutant of Trichoderma viride. Appl. Microbiol. 21, 152-154.
6. GRETHLEIN, H.F. and CONVERSE, A.O. (1982). Continuous acid hydrolysis for glucose and xylose production. Intern. Symp. Ethanol from Biomass, Winnipeg, Canada. Proceedings. Duckworth, D.E., Thompson, E.A., eds., 312-336.
7. BAILEY, M. and NEVALAINEN, H. (1981). Induction, isolation and testing of stable Trichoderma reesei mutants with improved production of solubilizing cellulase. Enzyme Microb. Technol. 3, 153-157.
8. CUSKEY et al. (1982). "Overproduction of cellulase - screening and selection", in Overproduction of microbial products. Krumphanzl, V. et al., eds., Academic Press, London, 405-416.
9. GALLO, B.J. (1981). Cellulase-producing microorganisms. U.S.Pat 4275163.
10. CHOUDHURY et al. (1980). Saccharification of sugar cane bagasse by an enzyme preparation from Cellulomonas: resistance to product inhibition. Biotechnol. Lett. 2, 427-428.
11. RICKARD, P.A.D. and LAUGHLIN, T.A. (1980). Detection and assay of xylanolytic enzymes in a Cellulomonas isolate. Biotechnol. Lett. 2, 363-368.
12. MEYER, H.-P. and HUMPHREY, A. (1982). Cellulase production by a wild and a new mutant strain of Thermomonospora sp. Biotechnol. Bioeng. 24, 1901-1904.
13. ENARI, T-M. et al. (1980). Comparison of cellulolytic enzymes from Trichoderma reesei and Aspergillus niger. Proc. Bioconversion and Biomed. Eng. Symp., New Delhi, India, Vol. I, 87-95.
14. TANGNU, S.K. et al. (1981). Enhanced production of cellulase, hemicellulase and β-glucosidase by Trichoderma reesei (RUT C-30). Biotechnol. Bioeng. 23, 1837-1849.
15. SADDLER, J.N. and BROWNELL, H.H. (1982). Pretreatment of wood cellulosics. Inter. Symp. Ethanol from Biomass, Winnipeg, Canada. Proceedings. Duckworth, D.E., Thompson, E.A., eds. 206-230.

16. RYU, D. et al. (1979). Studies on quantitative physiology of Trichoderma reesei with two-stage continuous culture for cellulase production. Biotechnol. Bioeng. 21, 1887-1903.
17. GHOSE, T.K. and SAHAI, V. (1979). Production of cellulases by Trichoderma reesei QM 9414 in fed-batch and continuous flow culture with cell recycle. Biotechnol. Bioeng. 22, 283-296.
18. ALLEN, A-L. and MORTENSEN, R.E. (1981). Production of cellulase from Trichoderma reesei in fed-batch fermentation from soluble carbon sources. Biotechnol. Bioeng. 23, 2641-2645.
19. HENDY, N. et al. (1982). Enhanced cellulase production using Solka floc in a fed-batch fermentation. Biotechnol. Lett. 4, 785-788.
20. LINKO, M. et al. (1977). Production of cellulases by Trichoderma viride in a semi-continuous fermentation process. 4th FEMS Symp., Vienna, Austria. Abstr. B40.
21. FREIN, E.M. et al. (1982). Cellulase production by Trichoderma reesei immobilized on κ-carrageenan. Biotechnol. Lett. 4, 287-292.
22. TAKAGI, M. et al. (1977). A method for production of alcohol directly from cellulose using cellulase and yeast. Proc. Bioconversion Symp., New Delhi, India, 551-571.
23. VIIKARI, L. et al. (1981). Hydrolysis of cellulose by Trichoderma reesei enzymes and simultaneous production of ethanol by Zymomonas sp. Proc. 6th Intern. Fermentation Symp., London, Canada, Vol. II, 137-142.

# STRAIN IMPROVEMENT FOR THE PRODUCTION OF MICROBIAL ENZYMES FOR BIOMASS CONVERSION

B. S. MONTENECOURT
Department of Biology and the Biotechnology Research Center
Lehigh University, Bethlehem, PA 18015, U.S.A.

## 1. INTRODUCTION

The international focus on cellulosic biomass as a potential for alleviating a part of the world food and energy needs has stimulated research in the area of saccharification of cellulose, hemicellulose and other plant polysaccharide materials. Hydrolysis of cellulose can be achieved through the action of a group of enzymes termed cellulases. Hemicellulose, which consists of polymers of five carbon sugars, is hydrolyzed by hemicellulases. The most predominant hemicellulase is xylanase. Glucans, which have both β1,3 and β1,4 linkages, are attacked by a group of hydrolytic enzymes generally termed β(1,3)(1,4) glucanases. Commercial cellulase preparations for the production of sugars, either for fuel or single cell protein, are prohibitive in cost due to the low yields and specific activities of the enzymes produced by the available microbial strains. Xylanases are also unavailable on a large scale since removal of the hemicellulose portion of the biomass may easily be achieved by dilute acid hydrolysis. β(1,3)(1,4) glucanases are used in the brewing industry in the malting process.

The production and utilization of all three types of enzymes suffer from the same physiological and biochemical controls. The synthesis of all three carbohydrases are subject to catabolite repression in the best representative microbial strains. In addition to poor productivity and specific activity, the enzymes themselves are subject to a variety of biochemical controls, the most important being that of end product inhibition. This end product inhibition results in inefficient saccharification of the substrate. Cellulose, and hemicellulose, may be hydrolyzed by acid. This route was extensively taken in the early part of the present century. However, acid hydrolysis suffers from a number of disadvantages. Although rapid, acid hydrolysis results in partial loss of the cellulosic and hemicellulosic material and the formation of toxic degradation products such as furfurol and hydroxymethylfurfurol. These degradation products would contaminate the sugar syrups and preclude the use of the sugars for either single cell protein production or high fructose syrups. Separation would be too costly a step for food production. Enzymatic hydrolysis offers a number of advantages over chemical hydrolysis. Enzymes are active at moderate temperatures and pressures. They show a high efficiency of conversion into a pure sugar product. In addition, highly toxic waste streams and high neutralization costs are avoided. With the advent of new hollow fiber and membrane technologies, the cellulase and hemicellulase enzymes could be recycled and reused or, alternatively, remain with the sugar product. In the application of cellulose hydrolysis to food production, the enzymes could serve as a source of protein. It is this author's understanding that *Trichoderma reesei*, one of the most prominant cellulolytic microorganisms, may soon be placed on the GRAS list and, at least in the United States, products from this microorganism will be permitted in food processing. The enzymes from Trichoderma are already allowed in food products in Japan. T. reesei is a good candidate for strain development. This microorganism produces cellulase, hemicellulase and β(1,4)(1,4) glucanases which could ultimately be used for processing biomass. The last enzymatic activity, β(1,3)(1,4) glucanase, is most often referred to as laminarinase since it attacks β1,3 polymers of glucose. A number of non-specific β(1,3)(1,4) glucanases also exist. In general, these enzymes will not attack cellulose. For purposes of clarity, I will refer to them as laminarinase-type activities to distinguish them from the cellulases.

At the present time cellulases and hemicellulases have been studied with respect to their application in alcohol fuel production. The laminarinase-type enzymes are employed in the brewing and malting industry to enhance malt recovery and filtration. They could, however, find application in increasing the digestibility of yeast and fungal cell walls in single cell protein production. In that all three enzymatic activities may be obtained from the same microorganism, T. reesei, I will use this example to illustrate how selection systems may be devised in order to identify mutant strains which have lost inherent control of enzyme production and may ultimately lead to their application in food and fuel production.

### 1.1 Cellulase enzymes

The synergistic activity of three different enzymes are required for the efficient hydrolysis of crystalline cellulose. Endoglucanases are the enzymes which initiate attack and act randomly to cleave internal β1,4 glycosidic bonds where the cellulose is amorphous (lacks a high degree of hydrogen bonding between the parallel cellulose chains). The second enzyme, cellobiohydrolase then cleaves cellobiose from the non-reducing termini to produce cellobiose. The end result of the two activities yields cellobiose and soluble cellodextrins up to $G_7$. The final enzymatic activity, cellobiase, then acts on the soluble products, cellobiose and cellodextrins, to yield glucose. The mechanism of this synergism is understood due to the contributions from a number of laboratories (1-6).

### 1.2 Hemicellulases

Hemicellulose consists of β1,4 linked D-xylopyranose chains which commonly contain α1,3 linked arabinofuranose and α1,2 D-glucanopyranose branches. Xylanase enzymes solubilize the xylan to xylobiose and oligomers of xylan. Although Trichoderma contains xylobiose, the hydrolysis rarely goes to completion and very little free xylose is usually produced. At least one of the endoglucanases of Trichoderma has been shown to be active on both carboxymethyl cellulose and xylan (7). Microbial xylanase has recently been reviewed (16).

### 1.3 β(1,3)(1,4) glucanases

Laminarinase-type enzymes are also produced by Trichoderma. The enzymes attack the β1,3 linkage of laminarin and do not attack cellulose. Laminarinases can be of both the exosplitting and endosplitting variety. Trichoderma seems to contain both types.

### 1.4 Physiological and biochemical controls

In developing selection systems for the detection of high yielding strains, it is important to understand the physiological and biochemical controls involved in the natural synthesis of these enzymes. In *Trichoderma* all of the aforementioned enzymes are extracellular glycoproteins. As they are inherently involved in carbohydrate metabolism, they are subject to catabolite repression. When the energy level of the cell is high, a condition which exists when readily metabolizable substrates are available, synthesis of cellulase, hemicellulase and laminarinase, is repressed. In addition, synthesis of each of these enzymes requires an inducer. Cellulose appears to efficiently induce all three activities (8,9). It is not as yet clear whether more efficient induction could be achieved if the natural inducers (xylan and laminarin) were present during growth.

All three types of enzymes are subject to end product inhibition. Within the cellulase complex, endoglucanase is inhibited by both glucose and cellobiose. Cellobiohydrolase is end product inhibited by cellobiose and to a lesser extent by glucose. Cellobiase is severely inhibited by glucose. In a saccharification situation where glucose, the desired product, builds up the inhibition occurs as a chain reaction. Glucose inhibits the cellobiase which results in the build up of cellobiose. Cellobiose, in turn, inhibits both the cellobiohydrolase and the endoglucanase. Both hemicellulases and laminarinases are end product inhibited in a similar fashion.

## 1.4 Selection systems

The selection systems for the isolation of high yielding microbial mutants which have been most successful are based upon developing a set of conditions in which the natural biochemical and physiological controls are operative. Initial mutants may be detected on the basis of clearing zone formation during growth on agar plates in which the insoluble biomass has been incorporated. However, clearing zone diameters will vary greatly with colony size and small differences of 5-10% are rarely detectable. One method of partially overcoming these difficulties is to restrict the colony size so that all of the potential mutants are forced to grow to the same colony diameter. This can be achieved by placing a plastic (2-3 mm thick) sheet over the agar and inoculating the potential mutants into the holes which have been drilled into the plastic. Each mutant can grow only to the internal diameter of the hole and cannot proceed further due to lack of oxygen. Since each mutant colony is now exactly the same size, one variable has been eliminated. The mutants can now be compared on the basis of their clearing zones. However, as mentioned previously, this method is useful only if there are large differences in the amount of enzyme being synthesized and small differences will not be detected.

A better approach is to devise a situation where the enzyme of interest is completely repressed. This can be achieved by the addition of a readily metabolizable substrate to the agar medium in addition to the insoluble inducer. Glycerol or glucose are both good catabolite repressors. This method has been successfully employed with *Trichoderma* in the selection of the high yielding strain Rut-NG14 (10). In this case, glycerol was added along with acid swollen cellulose. Selection was based upon clearing zone formation in the presence of the catabolite repressor.

In the selection of high yielding endoglucanase and xylanase mutants, the catabolite repressor can be combined with xylan or carboxymethylcellulose. Since neither of these substrates are opaque, a different selection system must be utilized. In this case a modification of the method of Teather and Wood (11) may be used. This method is based upon the selective precipitation of high molecular carbohydrate with Congo red dye. Areas adjacent to the colony where the carbohydrate has been digested will show zones of clearing. This method may be combined with restriction of the colony diameter to allow a definitive comparison of the various catabolite repression resistant mutants. In the selection of β(1,3)(1,4) glucanase mutants lichenin may be used as a substrate since laminarin is extremely expensive and plant β-glucans are difficult to prepare.

Selection for high yielding cellobiase and xylobiase strains can employ a number of techniques. It should be noted that, in Trichoderma, these enzymatic activities are partially constitutive and, in the case of cellobiase, about half of the activity is cell wall bound (8). Several selection systems have been devised for the isolation of cellobiase mutants (12). These include the use of esculin (6,7-dihydrocoumarin-6-glucoside) and ferric salts in the selection agar. Esculin is hydrolized to exculetin which reacts with the ferric salt to form a black precipitate. Methylumbelliferyl β-D-glucoside or p-nitrophenyl glucoside derivatives may also be employed. In the case of the umbelliferyl compounds, hydrolysis is detected by fluorescence. The p-nitrophenol product is, of course, yellow at alkaline pH conditions. Both the methylumbelliferyl and the p-nitrophenyl derivatives of xylose are commercially available although more expensive than the comparable glucose compounds. It should be noted that these are aryl substrates and not representative of the three disaccharide substrates.

End product inhibition of the cellulolytic enzymes could be relieved by the isolation of mutants which produce end product inhibition resistant enzymes. This approach has been taken in the isolation of mutants of T. reesei which produce an end product inhibition resistant β-glucosidase (12,13). An alternative avenue is the addition of exogeneous cellobiase which relieves the cascading inhibitory effect (13). For this approach to be plausible, the site of binding of the end product inhibitor cannot be the active site. However, non-competitive or uncompetitive inhibition has only been shown for the cellobiase of T. reesei (14) and the xylosidases of several other fungi (15) and may not be true for the enzymes which attack larger oligosaccharides. Since end product inhibitors will most likely function additionally as catabolite repressors, it would seem that this approach is worthy of attention in the selection of high yielding and end product inhibition resistant mutants for xylanase, endoglucanase and laminarinase.

Acknowledgements

This work was supported by the U. S. Department of Energy, Office of Basic Energy Sciences, Contract No. DE-AS05-80ER10702.000.

REFERENCES

1. PETTERSSON, G., FÄGERSTAM, L, BHIKHABHAI, R. and LEANDOER, K (1981). The cellulase complex of Trichoderma reesei. The Ekman-Days, vol. 3, p. 39-42. Stockholm.
2. ERIKSSON, K.E. (1981). Microbial degradation of cellulose and lignin. The Ekman-Days vol. 3, p.60-65. Stockholm.
3. WOOD, T.M. (1981). Enzyme interactions involved in fungal degradation of cellulosic materials. The Ekman-Days, vol. 3, p. 31-38. Stockholm.

4. RYU, D. and MANDELS, M. (1980). Cellulase: Biosynthesis and application. Enzyme Microbiol. Technol. 2:91-.
5. NISIZAWA, K (1973). Mode of action of cellulases. J. Ferment. Technol. 51:267-304.
6. REESE, E.T. (1977). Degradation of polymeric carbohydrates by microbial enzymes. Recent Advances in Phytochemistry. 11:311-367.
7. NHLAPO, D. Ph.D. Thesis, Rutgers University, New Brunswick, NJ.
8. MONTENECOURT, B.S., NHLAPO, S.D., TRIMINO-VAZQUEZ, H., CUSKEY, S., SHAMHART, D.H.J. and EVELEIGH, E. (1981). In Trends in the Biology of Fermentations for Fuels and Chemicals. (A. Hollaender, ed.) Plenum Press, New York. pp. 33-53.
9. TANGU, S.K., BLANCH, H.H. and WILKE, C.R. (1981). Enhanced production of cellulase, hemicellulase and β-glucosidase by Trichoderma reesei (RUT-C30). Biotechnol. Bioengineer. 23:1381-1396.
10. MONTENECOURT, B.S. and EVELEIGH, D.E. (1977). Preparation of mutants of Trichoderma reesei with enhanced cellulase production. Appl. Environment Microbiol. 34:777-787.
11. Teather, R. M. and WOOD, P.J. (1982). Use of Congo red - polysaccharide interactions in enumeration and characterization of cellulolytic bacteria from the bovine rumen. Appl. Environ. Microbiol. 43:777-780
12. CUSKEY, S.M., SHAMHART, D.H., CHASE, T., MONTENECOURT, B.S. and EVELEIGH, D.E. (1980). Screening for β-glucosidase mutants of Trichoderma reesei with resistance to end product inhibition. Develop. Indust. Microbiol. 21:471-480.
13. MONTENECOURT, B.S., CUSKEY, S.M., NHLAPO, S.D., TRIMINO-VASQUEZ, H., and EVELEIGH (1980). Strain development for the production of microbial cellulases. The Ekman-Days, vol 3, pp.43-50.
14. STERNBERG, D., VIJAYAKUMAR, P. and REESE, E.T. (1977). β-glucosidase; microbial production and its effect on enzymatic hydrolysis of cellulose. Can. J. Microbiol. 23:139-147.
15. GONG, C.S., LADISCH, M.R. and TSAO, G.T. (1977). Cellobiases from Trichoderma viride; purifications, properties, kinetics and mechanism. Biotechnol. Bioengineer. 19:959-981.
16. REILLY, P.J. (1981). Xylanases; structure and function. In Trends in Biology of fermentations for fuels and chemicals. (A. Hollander, ed.) Plenum Press, New York. pp. 111-131.

# MICROBIAL DELIGNIFICATION OF LIGNO-CELLULOSIC MATERIALS

P. ANDER and K.-E. ERIKSSON
Swedish Forest Products Research Laboratory
Box 5604, S-114 86 Stockholm, Sweden

## Summary

In this paper results which are relevant both for delignification of wood and for upgrading of agricultural wastes are discussed. Scanning and transmission electron microscopy are powerful techniques in the study of fungal growth and degradation of wood components. Results with cellulase-less mutants of white-rot fungi show that these can specifically delignify ligno-cellulosic materials although there still is a need for increase in delignification speed by the use of fungal strains with better properties. Such strains can be obtained by screening and mutation or by crossbreeding of monokaryotic strains.

## 1. INTRODUCTION

At the Swedish Forest Products Research Laboratory we have studied wood degradation by different wood-degrading fungi for about 15 years. These studies include both basic and applied research on cellulose and lignin degradation (1-3). One of the applied projects is called "Microbial delignification". In this project we have used cellulase-less mutants of white-rot fungi for specific deletion of lignin from wood chips in order to decrease the energy needed in mechanical pulping. Delignification of agricultural wastes such as straw and sugar-cane bagasse to increase the digestibility of agricultural waste for ruminants have been done in co-operation with other institutions.

Most mutation studies have been done with the white-rot fungus *Sporotrichum pulverulentum* which produces conidiospores in great amounts and has a high temperature optimum for growth (38-39°C). For the isolation of mutants, the spores have usually been treated with UV-light and spread on cellulose agar plates (4,5). From these agar plates fungal colonies which do not degrade cellulose can be isolated. The studies on delignification of wood chips with mutants obtained in this way have been divided into three parts:

a) Studies of growth conditions with different wild-type and mutant fungi (6)
b) Scanning and transmission electron microscopy studies (7,8)
c) Production of biomechanical pulp in laboratory scale (9).

### 1.1 Growth condition studies

The growth condition studies show that *S. pulverulentum* and its cellulase-less mutant Cel 44 grow 12 mm/day in birch, pine and spruce woods. A proportion of 10 % urea and 90 % $NH_4H_2PO_4$ was found to be optimal for growth and the best C:N ratio was 160:1 for pine and 200:1 for spruce. Two other mutants showed a somewhat higher optimum C:N ratio (6). Lignin determinations to show how different C:N ratios affect lignin degradation were not done in these experiments. Recent results (10) show that glucose impregnation represses cellulose degradation and stimulates lignin degradation. Malt extract can also be used as an additive to pine wood for specific delignification even by wild-type white-rot fungi (11).

1.2 Scanning and transmission electron microscopy studies

The electron microscopy (SEM and TEM) studies were performed to compare the growth pattern in wood of wild-type fungi and their cellulase-less mutants. In summary these studies (7) show that Cel 44 grows from one wood cell to another through already existing pits, pores and vessels. The wild-type fungi grow straight through the cell wall and also cause a thinning of the wood cell walls which was not found with the mutants. The TEM-technique (8) again shows that S. pulverulentum can grow through the fiber including the $S_2$ layer, while Cel 44 can not. The $S_3$ layer and the middle lamella are most resistant to microbial attack (8). The TEM-studies also show that lignin is modified at a distance of 2-3 µm from the hyphae by the wild-type. This indicates that lignin degradation does not necessarily require a close contact between the fungal cell and the lignin. If $^{14}CO_2$ is to be released from radioactive synthetic lignin in liquid cultures, the flasks cannot be shaken indicating that lignin in this case must be bound to the fungal cell before it is degraded (12). Future research on the primary attack on lignin, which perhaps involves the hydroxyl radical, will probably solve this discrepancy.

1.3 Production of biomechanical pulp

Most of the experiments with delignification of wood chips have been performed with Cel 44, which can degrade lignin without loss of cellulose under optimized laboratory conditions (5). Treatment of commercial wood chips of spruce and pine has been carried out in a bench-scale apparatus containing four 20 l steel cylinders (9). The best result is obtained with glucose-impregnated wood chips, whereas urea and $NH_4H_2PO_4$ impregnation probably leads to unwanted cellulose degradation although growth is good. The most favourable results show that paper made from glucose-impregnated Cel 44-treated spruce chips demanded about 20 % less energy input in the refining stage to obtain a value of 33 Nm/g in tensile strength as compared to untreated chips.

## 2. HOW TO DESIGN GOOD DELIGNIFYING MUTANTS?

Recently the final steps in lignin biodegradation have been elucidated (13) and we now know how vanillic acid is metabilized to β-ketoadipate and $CO_2$. Still further knowledge about the primary attack on lignin by enzymes or by hydrogen peroxide-derived hydroxyl radicals is necessary, before delignification of wood and agricultural wastes can be performed successfully (14). The mutants used so far have been cellulase-less but rather poor in phenol oxidase activity. This is probably a draw-back since evidence indicates that phenol oxidases are necessary in lignin degradation (15). Mutants which are better in this respect and also are xylanase-positive have now been found (16) and will be tested for delignification ability.

By crossing different interesting monokaryotic strains obtained from dikaryotic strains of S. pulverulentum (17) we will now try to obtain much better delignifying strains than the old ones. This will hopefully lead to better possibilities both for delignification of wood chips and for upgrading of agricultural wastes for use as cattle feed.

REFERENCES

1. ANDER, P. and ERIKSSON, K.-E. (1978). Lignin degradation and utilization by micro-organisms. In: M.J. Bull (ed). Progr. Ind. Microbiol., Vol. 14. Elsevier, Amsterdam, pp. 1-58.
2. ERIKSSON, K.-E. (1981). Swedish developments in biotechnology based on lignocellulosic materials. In: A. Fichter (ed). Adv. Biochem. Eng., Vol. 20 Bioenergy. Springer-Verlag, Berlin, Heidelberg, pp. 193-204.
3. ERIKSSON, K.-E. (1983). Recent advances on biodegradation of lignin. In: Boudet and Ranjeva, Journées Int. Group Polyphenols, Vol. II, Toulouse 1982. In press.
4. ERIKSSON, K.-E. and GOODELL, E.W. (1974). Pleiotropic mutants of the wood-rotting fungus *Polyporus adustus* lacking cellulase, mannanase and xylanase. Can. J. Microbiol. 20, 371-378.
5. ANDER, P. and ERIKSSON, K.-E. (1975). Influence of carbohydrates on lignin degradation by the white-rot fungus *Sporotrichum pulverulentum*. Sven. Papperstidn. 78, 643-652.
6. ERIKSSON, K.-E., GRÜNEWALD, A. and VALLANDER, L. (1980). Studies of growth conditions in wood for three white-rot fungi and their cellulase-less mutants. Biotechnol. Bioeng. 22, 363-376.
7. ERIKSSON, K.-E., GRÜNEWALD, A., NILSSON, T. and VALLANDER, L. (1980). A scanning electron microscopy study of the growth and attack on wood by three white-rot fungi and their cellulase-less mutants. Holzforschung 34, 207-213.
8. RUEL, K., BARNOUD, F. and ERIKSSON, K.-E. (1981). Micromorphological and ultrastructural aspects of spruce wood degradation by wild-type *Sporotrichum pulverulentum* and its cellulase-less mutant Cel 44. Holzforschung 35, 157-171.
9. ERIKSSON, K.-E. and VALLANDER, L. (1982). Properties of biomechanical pulp. Sven. Papperstidn. 85, R33-R38.
10. RUEL, K., BARNOUD, F. and ERIKSSON, K.-E. Ultrastructural aspects of wood degradation by *Sporotrichum pulverulentum*. Observations on spruce wood impregnated with glucose. To be published.
11. ANDER, P. and ERIKSSON, K.-E. (1977). Selective degradation of wood components by white-rot fungi. Physiol. Plant. 41, 239-248.
12. CHUA, M.G.S., CHOI, S. and KIRK, T.K. (1983). Mycelium binding and depolymerization of synthetic $^{14}C$-labelled lignin during decomposition by *Phanerochaete chrysosporium*. Holzforschung 37:2, in press.
13. ANDER, P., ERIKSSON, K.-E. and YU, H.-s. Vanillic acid metabolism by *Sporotrichum pulverulentum*: Evidence for demethoxylation before ring-cleavage. Arch. Microbiol. Under review.
14. ERIKSSON, K.-E. and KIRK, T.K. Biopulping and treatment of kraft bleaching effluents with white-rot fungi. Comprehensive Biotechnology, Vol. 3. In press.
15. ANDER, P. and ERIKSSON, K.-E. (1976). The importance of phenol oxidase activity in lignin degradation by the white-rot fungus *Sporotrichum pulverulentum*. Arch. Microbiol. 109, 1-8.
16. ERIKSSON, K.-E., JOHNSRUD, S.C. and VALLANDER, L. Degradation of lignin and lignin model compounds by various mutants of the white-rot fungus *Sporotrichum pulverulentum*. Arch. Microbiol. In press.
17. JOHNSRUD, S.C. Fruit body formation in the white-rot fungus *Sporotrichum pulverulentum*. To be published.

# SULFUR FREE PRETREATMENT OF LIGNOCELLULOSIC MATERIALS

U.P. GASCHE
Cellulose Attisholz AG
CH-4708 Luterbach
Switzerland

Summary

Recent development is presented of sulfur-free pulping processes for the manufacture of papermaking fibers from lignocellulosic material. Two main categories are outlined, the Organosolv-processes and alkaline pulping methods with delignifying additives. Delignification methods using organic solvents such as methanol, ethanol and phenols, with or without catalysts, do not render pulps of sufficient strength. For high strength characteristics, alkaline pulping is preferred. Methanol and the redox catalyst anthraquinone or combinations thereof may well replace the $S^{O}/SH^{-}$-redox system of the kraft process. Both chemicals accelerate the delignification rate and facilitate the dissolution of the native lignin by intensive fragmentation. Other additives such as amines and phenazine are not sufficiently effective. It is difficult to decide whether or not these new pulping methods could be of any use for bioconversion of lignocellulosic materials in the way of a sulfur-free pretreatment. It is hoped, however, that new ways of development in this field may be encouraged.

## 1. INTRODUCTION

Chemical pulping and pretreatment of lignocellulosic materials for bioconversion have one thing in common: The degradation or dissolution of the intercellular and intrinsic cementing substance lignin. In the case of biotechnical processes,the lignin content of wood and annual plant species acts restraining towards the penetration of carbohydrate degrading enzymes. In the case of paper making, residual lignin in the fibers impedes the formation of hydrogen bonds within the fiber web.

The outer cell wall of plant fibers, the middle lamella, forms a hard and hydrophobic sheating around the fiber. This lignin coat may, however, be partially destroyed by mechanical means, and the wood structure separated into single fibers, fiber bundles and fragments of fibers. A mechanical pulp resulting from such an operation is usually suited only for the production of low-grade papers such as newsprint. It is easily imaginable that, by analogy, a purely mechanical pretreatment of plant tissues leads in most cases to an unsatisfactory accesibility of microorganisms or added enzymes to the plant tissue carbohydrates and adequate action is hindered.

A pulp maker, trying to produce a fiber raw material of good quality for further processing, will always endeavor to achieve as extensive a delignification as possible. It will be his aim to do this quite selectively, though as to retain in the fibers as much of the native hemicelluloses as possible, which in turn are an important factor in the formation of physical strength during paper making. He will also try to retain the cellulose component of the wood more or less as a whole. The extensive cellulose molecules, which are aggregated to micelles within the cell wall, impart the necessary strength to the fiber and with it to the final paper product. Therefore, a most gentle delignification is aimed at. This holds true for all pulps produced from wood and annual plants, with the exception of chemical pulps to be further processed to cellulose derivates. However, this latter portion amounts to only 4.5 % of the worlds total pulp production.

In reality, it is inevitable to hydrolize large portions of the hemicelluloses during delignification. Quite early in pulping history it was realized that these carbohydrates are available mostly as mono-sugars after a digestion in an acid medium. As a result, alcoholic fermentation is practiced on an industrial scale since 70 years and a yeast fermentation since 40 years, thus utilizing the mono-sugars in spent sulfite liquor by means of biotechnical processes. Actually, this experience could be put to use by bio-technologists, which means that after an intensive pretreatment of lignocellulosic material the enzymatic degradation and bioconversion of the highly carbohydrate enriched material could follow.

In the context of bioconversion, apparently such a chemical pretreatment stage is too far-reaching. To avoid the relatively large application of chemicals and energy, the biotechnologist prefers to employ microorganisms and enzymes to decompose and further utilize the chemical main components of plant cells in single-stage or multi-stage processes.

The pulp maker, on the other side, is not too happy either with the world-wide most commonly used acid or alkaline pulping processes. In both cases, he requires selective acting chemicals to achieve a chemical decomposition of the lignin. Such chemicals are, without exception, sulfur containing. During pulping operations, approximately 50 % of the wood substance is being dissolved. With conventional utilization technologies, the predominant part of this dissolved portion is being burned, and only a small fraction of the lignin derivates and carbohydrate fragments may find use in further processing. It is during this burning process, and even during preceeding process stages that emissions of sulfurous gases occur, which in turn may burden the environment.

This is the main reason why there are intensive research efforts under way already for a number of years, to find a new basic method of delignification for wood and annual plants. Researchers not only endeavor to replace the more than 100 year old pulping processes, but also try to develop a process, in which delignification eventually may be completed in the pulping stage instead of both in the pulping and the bleaching stages. The following short survey of the activities in this field may, perhaps help to offer new ideas with regard to a sulfur free treatment of lignocellulosic materials also for bioconversion.

## 1. SOLVENT PULPING

Delignification of hardwoods and softwoods may be achieved to a large degree using 1:1-solutions of ethanol and water at temperatures of 185 °C to 200 °C, according to Kleinert (1). Here, it is apparently essential that the reaction temperature is reached very quickly. It is further reported that this ethanol-water pulping is most advantageously operated as a continuous counter-current process. Dissolved components and liberated organic acids may, thus, be removed from the system without considerable time of direct contact with the reaction material. Hydrolysis of the hemicelluloses is subdued, which in turn has a favorable effect on the physical strength properties of the pulp. The recovery of the ethanol phase is achieved by flash evaporation. Semi-industrial trials have shown that total alcohol losses may be kept, apparently, in the range of 1 % as based on pulp.

By stripping the ethanol from the waste digestion liquor, the dissolved lignin components become insoluble in the remaining aqueous liquid. They precipitate as a quasi-melted specific heavy phase. Wood sugars and other hydrolysis products of the hemicelluloses remain in solution. Small quantities of phenolic lignin decomposition products are removed from the supernatant waste liquor by an active carbon treatment. Consequently, the waste liquor may be used as a feedstuff or a feedstuff raw material.

The resulting pulp, however, leaves much to be desired. Even though yields are in the range of sulfite pulps, physical strength properties reach in no way the level of sulfate pulps, which would be a very desirable basic requirement to introduce such a new pulp grade on the market. As it is, there is little sense in marketing a product, which has hardly a chance to succeed for reasons of quality.

In the mid-sixties, we have had similar experiences during experimental trials, using an aqueous methanol solution and an acid catalyst to achieve a delignification of lignocellulosic material. The resulting pulp quality was comparable to that of a sulfite pulp. In spite of these facts, a patent was issued early in the nineteen eighties, which describes the pulping of lignocellulose with aqueous alcohol-catalyst mixtures (2). The catalyst salt is a sulfate, nitrate or chloride of calcium, magnesium or barium. The preferred organic solvent is methanol, and the catalyst is applied with or without the addition of minor quantities of strong mineral acids.

Once again, these research findings confirm that alcohols possess delignifying properties in an aqueous medium and at high reaction temperatures. Overall, these and similar possibilities are summed up under the general term ORGANOSOLV-Processes. For reasons already mentioned, these sulfur-free pulping processes will probably not gain to much importance. A biotechnologist could, perhaps, perceive the one or the other application possibility of this process technology for a chemical pretreatment of biomass or for the fractionation of biomass.

To this end, a Swiss research institute (3) has done a lot of work and developed the Phenol-Process, which claims upgrading of all three main components of lignocellulosic materials without expensive chemicals.

Delignification of wood using phenol and phenol mixtures was already investigated by Schweers (4) some time ago. Contrary to the Battelle-Process however, temperatures at approximately 160°C and hydrochloric acid as a catalyst had to be applied. Nevertheless, pulps produced by either process will never be in a position to compete seriously with Kraft pulps, alone from the aspect of their physical properties.

## 2. ALKALI-METHANOL PULPING

It is a well known fact that fibrous materials derived from wood or annual plant species exhibit the most favorable strength properties after a delignification in a strongly alkaline medium. However, an aqueous solution of sodium hydroxide alone is not sufficient, and a "buffering agent" is needed to delignify fibers as selectively as possible. In the Kraft process, this function is fulfilled by the sodium sulfide component of the cooking liquor.

A replacement of this sulfur compound presents itself in the heretofore discussed application possibilities of the organic solvent methanol as a relatively cheap delignification agent. We have indeed produced very acceptable pulps on a laboratory scale showing good quality characteristics. The reason why methanol remains in the foreground of our endeavors is the fact that during chemical digestion of wood in a strong alkaline medium methanol is formed by cleavage of the methyl ether groups of the lignin. This in turn helps to replenish methanol losses, which will occur to a certain extent during chemical recovery.

Delignification trials using spruce wood were conducted at constant active alkali levels and varying methanol concentrations in the cooking liquor. The following results were obtained (Fig. 1) :

- A most selective delignification is achieved at a methanol concentration of 40 - 50 %.
- Without the addition of methanol cooking stagnates at a relatively high residual lignin content, as is well known (soda pulp).

- Likewise, pure undiluted methanol-sodium hydroxide mixtures without water also result in an incomplete delignification with considerable portions of rejects.
- The optimal methanol concentration in the cooking liquor leads, not only, to the lowest residual lignin content, but also, to the highest fiber yield, unbleached as well as bleached, i.e. also the final lignin-free fiber yield is at its optimum.

It was also established that most satisfactory physical strength properties were obtained at methanol concentrations in the range of approximately 40 %. It seems, therefore that this system is basically suited for the delignification of wood and annual plants.

## 3. ALKALI-METHANOL PULPING WITH CATALYSTS

The pulping chemist attempts to achieve in the fibers as low a residual lignin content as possible by means of digestion,reducing thereby the environmental effluent load of the bleach plant. Using methanol as a delignifying agent in an alkaline medium, residual lignin contents in the range of only 6 % are attainable. A reduction of the residual lignin to lower values leads to undesirable carbohydrate losses and to impaired physical strength properties. It is therefore understandable that, worldwide, great activities were and still are underway to discover a chemical compound, which would possess a specific and selective delignifying effect, a chemical that could replace the sulfurous components in the cooking liquor of the classical pulping processes (Sulfite and Kraft) when added in small quantities.

Of the many thousand chemical compounds, which were altogether investigated, only anthraquinone (AQ) seemed fo fulfill the required criteria in a satisfactory way.

In 1976, Holton (5) discovered AQ to be an effective additive in alkaline pulping. AQ is insoluble in water and only slightly soluble in organic solvents. Most scientists consider AQ functioning as a redox catalyst. It is assumed and meanwhile confirmed that carbohydrate end groups are oxidized by AQ. Lignin then completes the catalytic cycle by oxidizing anthrahydroquinone (AHQ) back to AQ (Fig. 2)

Even very small quantities of AQ are, in actual fact, capable of intensifying the delignification process of spruce and beech. Additions of 2 kg AQ per ton of spruce wood and 1 kg AQ per ton of beech wood are already sufficient to further reduce the residual lignin content in the system NaOH - $CH_3OH$ (Fig. 3)

To obtain a more comprehensive survey of the effectiveness of both additives, methanol and AQ, we have conducted a series of standard cooks at conditions as described in Fig. 4. With pure Soda-Process cooks (NaOH), even an increased application of active alkali results invariably in lignin-rich pulps, which can not be bleached. An addition of 40 % W/W methanol already promotes delignification considerably. An even more favorable effect shows, without doubt, AQ as an additive.

This becomes most evident if 2 kg AQ/t of wood are added to the NaOH - $CH_3OH$ system, which effects an additional increase in the rate of delignification. At the same time, it was observed that methanol as well as AQ exert a stabilizing effect on certain hemicellulose fractions.

We were able to establish that a total fiber yield of 53.8 % at a residual lignin content of only 4.8 % is quite obtainable, applying the NaOH - $CH_3OH$ - AQ system at an active alkali concentration of 21 %. Experimental cooks with the NaOH-AQ system resulted in total fiber yields of 52.5 % at a residual lignin content of 6.1 %, whereas the NaOH - $CH_3OH$ variant resulted in 52.3 % at 7.6 % residual lignin. This means that we obtained lignin-free fiber yields in the three before mentioned cases of 51.2 %, 49.3 % and 48.3 % respectively.

4. ALKALI PULPING WITH AQ

The investigations just described show clearly that the catalyst AQ has a favorable effect on the rate of delignification. The investigated effects were, however, confined to the final phase of the pulping reaction and to characterizing the properties of the pulps obtained therefrom. In the following, we studied to what extent the accelerating effect of the AQ/AHQ-reaction continues during the entire digestion of wood, or if it takes place only at a certain temperature range.

In Fig. 5, the course of delignification is shown with and without the addition of AQ. It seems that the beneficial effects of AQ become active already at low temperatures. Both delignification curves presented in Fig. 5 were derived from cooks under identical digesting conditions, using spruce wood chips of defined size (2 x 1,5 x 0.3 mm).The course of delignification of the two types of cooks deviates considerably. Also Gierer (6) confirmed recently the beneficial effect of the AQ/AHQ redox system by alkaline delignification experiments in the presence of AQ using wood shavings from Pinus sylvestris. Apparently, this effect manifests itself, not only, in an accelerated delignification rate, but also, in stabilizing carbohydrates.

However, the oxidative alterations of the carbohydrates seem not to be restricted to the end groups alone. It appears that the anhydro sugar units are, additionally, provided with carbonyl groups, which are added at the 2, 3 and/or 6 positions of these. This in turn increases the sensitivity of the polysaccharides towards alkaline degradation. In spite of the fact that delignification in a NaOH - AQ medium is more selective as is the case in the Kraft-Process (NaOH - $S^{o}/HS^{-}$), it may be due to this side effect that physical strength properties of the fibers derived from the former process do not quite correspond to those of the latter (6).

5. CHARACTERIZATION OF SPENT LIQUORS RESULTING FROM THE VARIOUS PULPING PROCESSES

The components of spent liquors are composed of, essentially, inorganic salts, organic acids, which were formed from degraded hemicellulose sugars, and of fragmentized lignin. The latter constitutes the main portion of the organic component. We were therefore most interested in the molecular weight distribution of the spent liquor lignin resulting from the digestion of wood in the systems NaOH - $CH_3OH$ and NaOH - $CH_3OH$-AQ respectively. As a comparative standard substance we used spent sulfite liquor. The delignification degree (Kappa-Number) of the three pulps was kept within the same range (40-45 K.N.), from which the three spent liquors were taken for analyses. The tried and well proven gel-chromatographic method was used to determine the molecular weight distribution curves, using Sephadex-Dextran gels. Our calibration standard was Polyfon O.

The spent sulfite liquor clearly stands out against the alkaline spent liquors. The molecular weight range of over 60 % of its lignin components is well over 10,000 D, and its average molecular weight was found to be approximately 15,000 D. Delignification using methanol, on the other side, results in a dissolved wood substance of considerable lower molecular weight, averaging at approximately 10,000 D. The molecular weight portion above 10,000 D was estimated at 32 %. The smallest percentage of high-molecular weight components was found in the spent liquor derived from wood digestion with an AQ addition (Fig. 6).

These findings lead to the conclusion that, apparently, the delignifying catalyst also promotes fragmentizing of the native lignins, which again helps to increase the rate of delignification. It is therefore logically consistent that the average molecular weight of the AQ-containing spent liquor was found to be 8,500 D, and, in comparison, that of the AQ-free NaOH - $CH_3OH$ spent liquor to be 10,000 D.

## 6. OTHER POSSIBILITIES OF SULFUR-FREE PULPING

Within the scope of this review I only like to mention the nitrogenous redox catalysts, the investigations of which were described by Fleming and others (7). Phenazine for instance furthers the delignification rate of black spruce in soda liquor, but with only one third the efficiency of AQ. This and other reasons make the commercial application of phenazine and similar compounds unlikely.

The same holds true for amine pulping. Organic amines also show a delignifying effect when added to soda cooks at a rate of 10 - 50 % (8). Amines such as monoethanolamine or ethylenediamine improve delignification. A stabilizing effect on the carbohydrates, however, could not be observed. Pulps produced with amine additives have shown a remarkably high tearing strength, but, at the same time, a lower tensile strength.

## 7. CONCLUSION

Sulfur-free delignification processes as discussed in the foregoing have only then a chance to replace the worldwide dominant Kraft process, if researchers succeed to produce pulps of equal quality to that of Kraft pulps. Alkaline pulping processes applying suitable sulfur substitutes as additives fulfill this requirement to a large extent. A high delignification degree may be obtained at relatively gentle conditions with the three-component system NaOH - $CH_3OH$ - AQ. Since methanol as well as anthraquinone exert a stabilizing effect on certain hemicelluloses, a higher lignin-free fiber yield results, as may be obtained with the Kraft process. This higher yield must, however, compensate for the unavoidable methanol losses. The multicomponent pulping process should, wherever possible, be operated as a continuous process.

It is doubtful for economical reasons that these alkaline multicomponent systems are suitable for a sulfur-free pretreatment of lignocellulosic materials, with the exception perhaps that the dissolved lignins and hemicelluloses are up-graded to products of higher value with the help of biological processes.

The Organosolv processes will hardly reach any significance as a commercial pulping process. It is quite conceivable though that they may be suited as a pretreatment stage in bioconversion of lignocellulosic materials, since they permit, to a certain extent, a fractionation of the three main components lignin, hemicelluloses and cellulose, without a simultaneous chemical modification. Nevertheless, economic border conditions may prove to be prohibitive.

REFERENCES

1. KLEINERT, T.N., Das Papier 30 (10A): V18 (1976)
2. PASZNER, L. and PEI-CHING CHANG, UK Pat. Appl. GB 2040332 A (August 28, 1980)
3. N.N., Europ. Chem. News 4 (3), 1 (1982)
4. SCHWEERS W. and RECHY M., Das Papier 26 (10A), 585, (1972)
5. HOLTON H.H., Can. Pulp Pap. Ind. Tech. Mtg. (1976), February, A 107
6. GIERER J., KJELLMAN M. and NOREN I., Holzforschung 37 (H1): 17 (1983)
7. FLEMING B.I., KUBES G.J., MACLEOD J.M. and BOLKER H.I., Tappi Pulping Conference, Sept. 1978
8. FLEMING B.K., KUBES G.J., MACLeod J.M. and BOLKER H.I., Tappi 62 (6) : 57 (1979)

ACKNOWLEDGMENT

The author is indebted to Schweizerischer Nationalfonds zur Förderung der wissenschaftlichen Forschung for financial support and to Mr. C.S. Weber for translating this paper into English. The assistance and the skillful experimental work of Messr. F. Binder and R. Stampfli is gratefully acknowledged.

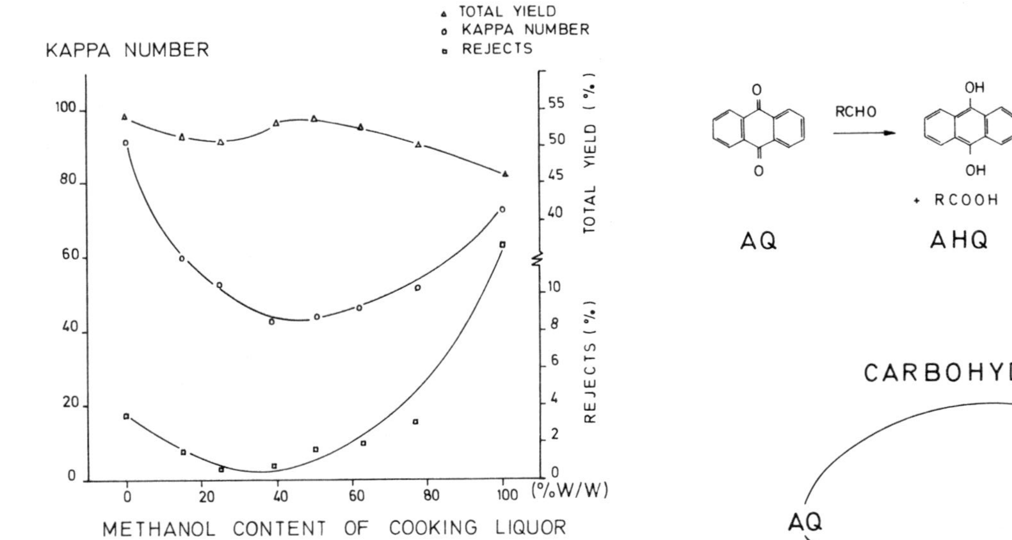


Fig.1 Kappa number, total yield and rejects of unbleached pulp as a function of the methanol content of the cooking liquor at constant cooking conditions. (22 % active alkali, wood/liquor ratio 1:3.5, max. temperature 160° C, time at max. temperature 3.5 hours.)

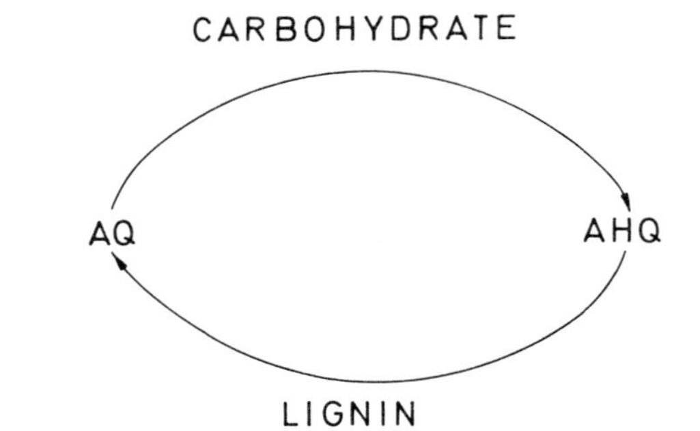


Fig. 2 MECHANISM OF ANTHRAQUINONE PULPING

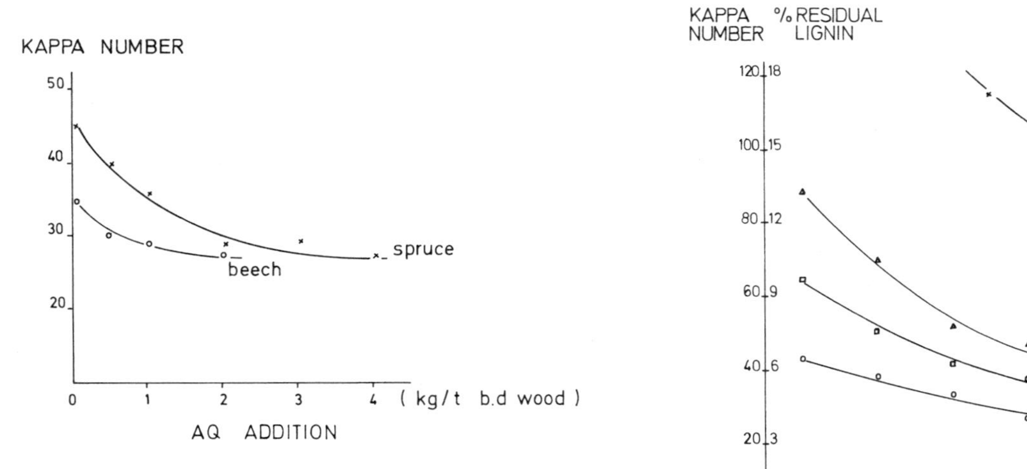


Fig.3 The effect of AQ on NaOH / $CH_3OH$ - delignification of spruce and beech wood.

Cooking conditions: 3.5 hours at 160°C, 22 % active alkali and 2 hours at 160°C, 18 % active alkali respectively (Percentage AA based on wood).

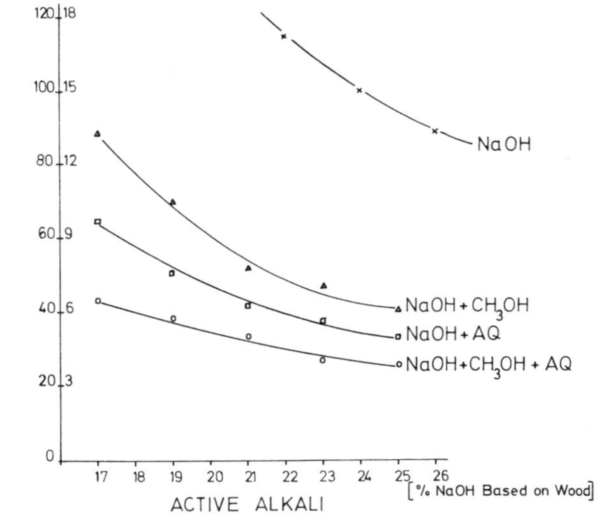


Fig.4 The effect of varying active alkali charges on the residual lignin content of unbleached pulp cooked at constant conditions of temperature and reaction time (3.5 hours at 160°C) in combination with methanol and AQ.

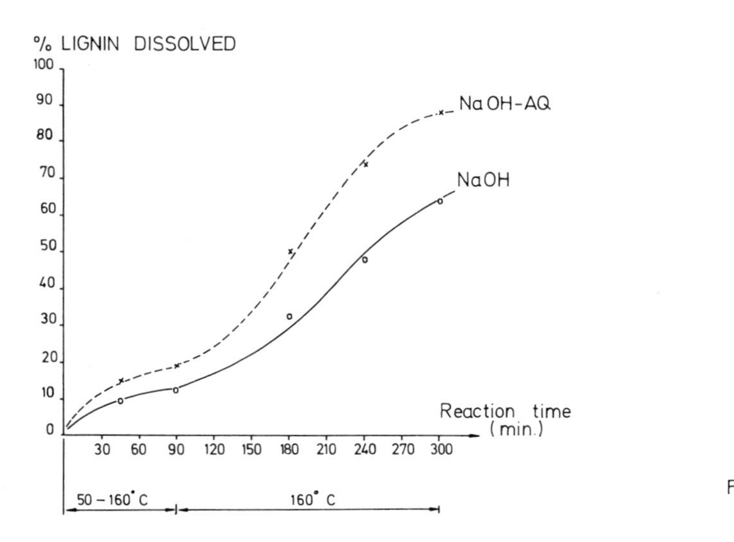


Fig. 5 Alkaline delignification of spruce wood chips, with and without AQ addition ( 2 kg/t of wood.)

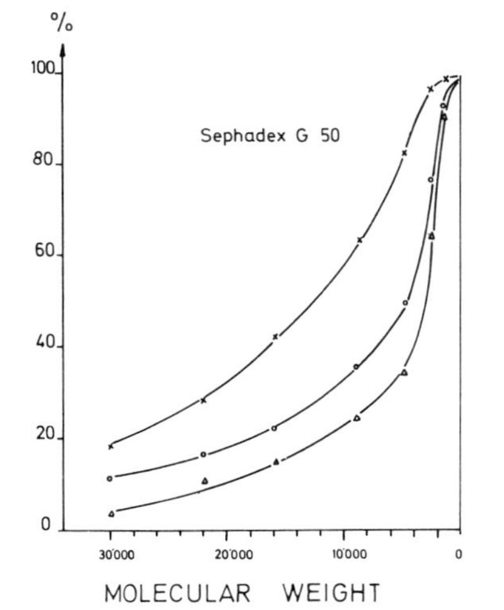


Fig. 6 Molecular weight distribution of spruce spent liquors.
( Graphical presentation of summarized molecular weights.)

× Ca - Sulfite
∘ NaOH-$CH_3OH$
△ NaOH-$CH_3OH$-AQ

## SUBJECT AREA 1

## PRODUCTION OF SCP ENRICHED SUBSTRATES FROM CELLULOSIC MATERIALS

Chairman : A. FIECHTER

Review paper :
Cellulases: delicate exoproteins - Demonstration of multi-enzyme complexes within the culture fluid of Trichoderma reesei

### Lignin and Lignocellulose

Factors determining lignin decomposition and in vitro digestibility of wheat straw during solid state fermentation with white rot fungi

Solid culture using alkali treated straw and cellulolytic fungi

Studies on the extracellular cellulolytic enzyme system of Chaetomium cellulolyticum

### Process Development

Pre-treatment and conversion of straw into protein in a solid-state culture

Pre-treatment of cereal straws and poor quality hays

Production of mycelial biomass on waste water in a rotating disc fermenter

Protein enrichment of pretreated lignocellulosic materials by fungal fermentation

### Carbohydrates

Protein enrichment of starchy materials by solid state fermentation

Solid state fermentation of cassava with Rhizopus oligosporus NRRL 2710

Utilisation de la bagasse traitée par la soude pour la production de protéines d'organismes unicellulaires

Conversion of agricultural and industrial wastes for cellulose hydrolysis

Cellulose hydrolysis of papermill sludge

Protein enrichment of sugar beet pulps by solid state fermentation

# CELLULASES: DELICATE EXOPROTEINS

## Demonstration of multi-enzyme complexes within the culture fluid of Trichoderma reesei

B. Sprey and C. Lambert
Institut für Biotechnologie der Kernforschungsanlage Jülich

Summary

Based on different preparative isolation procedures (bi-directional gradient PAGE, flat-bed IEF) the complete exoprotein composition of Trichoderma reesei as present in the culture fluid was monitored by 7 (8) different enzyme tests. The rationale of this study was to find out to what extent purity in terms of homogeneous fractions - obtained by the different purification procedures - is a criterion for homogenity of an enzyme isolated. By means of immunology and titration curve techniques, it could be demonstrated that homogenity of a fraction - independent of the purification method used - reflects more the purity of exoproteins complexes with apparent molecular weights and apparent isoelectric points than that of single proteins.

## 1. Introduction

Cell-free fungal hydrolases, especially cellulases, have been frequently subjected to reviews (1-4). It seems to the authors that a further reviewing would not help to overcome the obvious bottleneck of progress in this field when compared to the amylase-technology, a topic fulfilling both the requirements of application and intimate knowledge of the (sequenced) amylase protein structure and its function (5).

The lack of information on the true state of enzymes, such as cellulases and xylanases in fungal culture fluids, seems to be such a bottleneck. Conflicting results - such as broad variances of molecular weights obtained, of pI, of products formed, can in part be explained by the current concept of purifying, treating and interpreting exoenzymes as single proteins. In contrast to this conceipt and based on experiences of evaluating the complete culture fluid of Trichoderma reesei with different biochemical purification procedures, we feel that the real state of exoenzymes is that of being involved in (multi)-enzyme complexes when working under detergent-free conditions.

Cellulases and xylanases acting on water-insoluble substrates generally meet some criteria that make this non-homogeneous catalysis extremely difficult for definitions in terms of enzymology. Difficulties arising from this subject mainly cover the following sources:

- differences and inhomogenities in the composition of the substrates used (e.g. the use of non-natural and artificial cellulose sources as Avicel, carboxymethylcellulose CMC, filter-paper F.P., hydroxy ethylcellulose HEC, dyed Avicel etc.)
- difficulties derived from definition and characterization of 'endo- and exo'cellulases acting on different cellulose sources ('Avicel'ase, 'CMC'ase, 'F.P.'ase) together with using the liberation of reducing sugars as single criterion. These difficulties generally can't even be overcome by standardized cellulase tests (6)
- method-derived difficulties to separate and characterize non-soluble, higher-ranked cello- and xylooligodextrins occuring with the polymer degradation
- limited value of protein purification procedures valid for purification of cytosol proteins (e.g. gel filtration, affinity chromatography, ion-exchange chromatography)

Moreover, celluolytic enzyme action appears even more complex with regard to the following parameters

- introduction of endo- and exo-wise acting cellulases (7, 28)
- synergistic effects (8)
- adsorption phenomena between polymeric substrates and enzyme complexes (9)
- inhibitory effects of endproducts formed (10)
- an unknown mode of glycosilation of exoproteins complexes, introducing apparent pIs, thus making a comparison of cell-free exo-proteins complexes from different species hardly feasible (11)

Thus in contrast to the tremendous, detailed information on

enzymatic cellulysis gathered(c.f. 12),the composition and action of cellulolytic exoproteins is - to the author's ideas and experiences - a still poorly understood phenomenon. A fact often overlooked when working with exoproteins of Trichoderma is the presence of exoprotein complexes in culture fluids.

Biochemical purification procedures for both cell-free cellulases and xylanases from Trichoderma often follow prescriptions valid for cytosol enzymes, e.g. demonstration of an n-fold purification within a given scheme (e.g. selective ammonium sulfate precipitation, different gel chromatographic techniques etc.). It should be noted that in the case of cell-free cellulases these steps are valuable for a pre-purification to achieve purity of exo-enzyme complexes but not of a single enzyme, when working under detergent-free conditions.
Demonstration of reasons for the more or less exclusive distribution of enzyme complexes in the culture fluid of Trichoderma are the subject of this article and biochemical data to corroburate this conceipt will be given in this contribution.

The multi-enzyme character of exoprotein complexes in mind, the conclusions derived from the purification procedures currently used can be misleading and highly erraneous, when referring one enzyme activity to one homogeneous fraction only. Both the multi-enzyme conceipt of exoproteins and the occurence of lipases, nucleases, proteases, laminaranases, mannanases, chitinases, xylanases, xylosidases, β-glucosidases, amylases (12) and probably additional hydrolases in the culture fluid, testing for purity always should - as a logical must - include several tests. Thus, this conceipt requires the consideration of the occurence of additional enzyme activities in homogeneous fractions not as impurities, contaminants etc., but as the possibility of this enzyme involved in the multi-enzyme complex purified.

The intention of this study was to demonstrate enzymatic heterogenity in complex composition of homogeneous, single-banding fractions of exoproteins of T. reesei after preparative polyacrylamide gel electrophoresis (PAGE) or preparative iso-electric focusing (IEF). The prerequisites for the following experiments were:

- use of several, different enzyme tests ('endo-, exo'cellulases, β-glucosidase, xylanase, xylosidase, protease)
- testing the complete culture fluid composition of T. reesei QM 9414 (ATCC 26921), grown under detergent-free conditions
- use of separation and preparation methods with high-resolution, i.e. bi-directional gradient PAGE and flat-bed IEF
- assay of conventionally raised antibodies against homogeneous complexes obtained by preparative PAGE and IEF
- adopting the titration curve technique on pure exoprotein fractions together with a lay-on zymogram technique for the identification of single enzymes present in the multi-enzyme complex

## 2. Materials and methods

### 2.1 Preparative polyacrylamide gel electrophoresis

Exoprotein complexes were separated from lyophilized culture fluid of Trichoderma reesei QM 9414 (ATCC 29921) grown on 1 % (w/v) Solca Floc as carbon source. For separation (30 - 100 mg protein) bi-directional techniques of slab gradient polyacrylamide gel electrophoresis (PAGE) were used. Samples were separated in 20 x 20 cm slabs of 1.5, 3.0 or 5.0 mm thickness. Buffers used were Tris-HCl, pH 8.8, or a β-alanine-KOH, pH 4.3 ( 13, 14). Separation conditions depending on gel thickness were: 10 - 25 mA, 100 - 300 V for 18 h. Detection of the protein complexes in the PAGE slabs achieved by a blotting technique using Whatman No I filter paper or Polygram CeL 400 sheets (20 x 20 cm, Macherey and Nagel, Düren) coated with microcrystalline cellulose Avicel as template. Cellulose-containing sheets were gently pressed to the surface of the slab gels (15 min.) removed and subsequently hot-air dried with a fan. Proteins present in the blot were fixed in Coomassie-stain solution (as routinely used for protein staining) for 15 min.. Cellulose sheets were destained with 2-propanol, acetic acid,

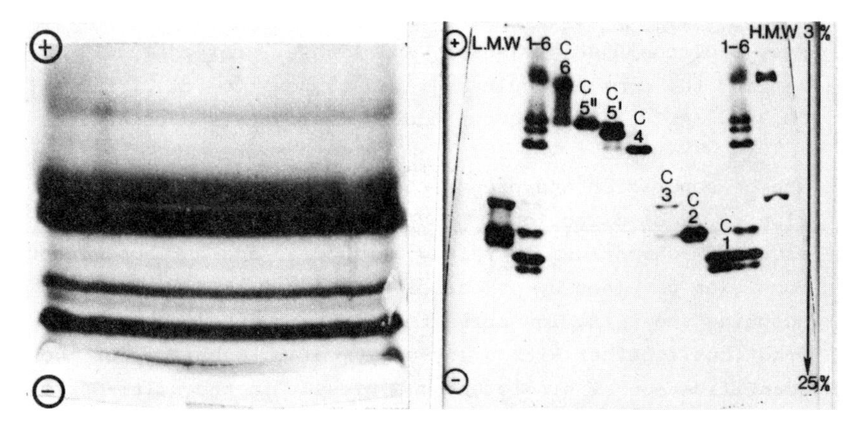


Fig. 1: Coomassie-stained blots (Polygram Cel 400 cellulose foils) from Trichoderma culture fluid separated by gradient PAGE in the cathodic direction (β-alanine-KOH, pH 4.3). Preparatively isolated exoproteins (C1-C6) by the cellulose-blotting technique are given in Fig. 2.

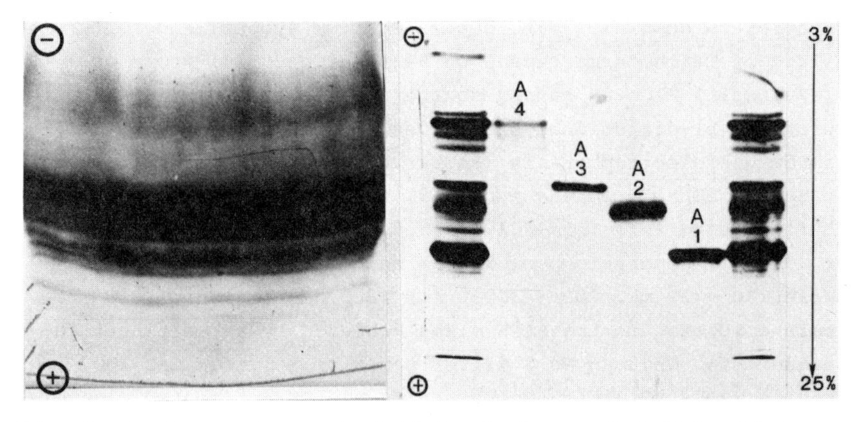


Fig. 3, 4: Coomassie-stained blots (Polygram Cel 400 cellulose foils) from Trichoderma culture fluids separated by gradient PAGE in the andodic direction (Tris-HCl, pH 8.8). Preparatively isolated exoproteins complexes (A1-A4) by the cellulose-blotting technique are given in Fig. 4.

$H_2O$ (25:10:65 v/v/v,15 min.), washed in 15 % acetic acid (v/v, 15 min.) and hot-air dried (Fig. 1, 3). Protein complexes in the gel were cut out with a scalpel by underlaying the slab gel with the Coomassie-stained blot and illuminating the gel from below. Cut-out gel strips were filled in a X-press (Enerpac, Applied Power, Sweden), frozen at -20°C and minced with 8-cycles. Proteins were eluted from gel fine particles by stirring in a 3-5-fold excess of distilled water under stirring overnight (4°C). Gel fine particles were removed by ultracentrifugation (120,000 g, 1 h), and repeatedly (2-3 times) water-washed. The supernatants were desalted and concentrated with $N_2$-pressure (Amicon UM 0.5) ultrafiltration.

### 2.2 Preparative flat-bed isoelectric focusing

Preparative IEF in flat-beds was followed according to the instructions given by Radola (15) using granulated gels (Ultrodex, LKB) and Servalyte (pH 2-5, 3-6, and 6-8) as carrier and ampholyte, respectively. Separation and detection methods of proteins in the gel beds were those given previously (11). Refocusing of protein complexes were performed in ultrathin ampholyte-acrylamide gels (16).

### 2.3 Protein titration curves and zymograms

pH-mobility curves of protein complexes were obtained using the method of Righetti and colleagues (17), under urea- and urea-octylglucoside conditions reported earlier (11).

### 2.4 Enzyme assays

Cellulase (E.C. 3.2.1.91) activity was tested towards Avicel, carboxymethylcellulase CMC, dewaxed cotton, and filter paper F.P. as substrate under conditions given by Berghem and Pettersson (18). Activities were expressed as units (u) of released reducing sugars (19). β-glucosidase (E.C. 3.2.21) activities were determined with p-nitrophenyl-β-D-glucoside or cellobiose as substrate (18). F.P. activity was estimated following Mandel's procedure (20). Proteinase activities were tested according to (21). Xylanase and xylosidase activities were assayed as reported earlier (11).

## 2.5 Immunological techniques

Antibodies were raised to purified exoenzyme complexes preparatively obtained with PAGE, following the prescription given by (22). Rabbits were immunised 3 times at weekly intervals with 1 mg of purified protein each. Sera were purified for IGG (23). Detection of immunopreciptates was achieved by diffusion of purified antibodies in the Ouchterlony double diffusion test in 1.5 % agarose plates (Tris-borate, pH 8.3). Immunoelectrophoresis in agarose plates (Corning System ACJ) was performed using both a Tris-borate buffer, pH 8.8, in the anodic direction or a β-alanine-KOH system, pH 4.3, for separations performed in the cathodic direction. Antisera were filled into troughs (10 μl) sidewise to the electrophoretic separation. When the acidic separation system was used, agarose plates had to be equilibrated with Tris-borate buffer, 3 min., to prevent precipitation of IGG in acidic milieu.

# 3. Results and discussion

## 3.1 Characteristics of exoprotein complexes preparatively purified from cathodic separations with PAGE (β-alanine-KOH-system, pH 4.3)

Fig. 1 represents 7 exoprotein complexes preparatively isolated by the blotting technique given under Materials and methods. Though cut-out from gel-strips containing homogeneous Coomassie-positive bands (visualized in the corresponding cellulose blots) fraction C3, C5', and C6 tend to form additional minor bands upon reelectrophoresis, positioned at same height of C2 and C5", respectively (Fig. 2). Corresponding enzyme activities of β-glucosidase and cellulase are given in Table 1. A qualititative estimate of the cellooligomers, cellobiose and glucose based upon sugar chromatography is shown in Table 2. Fractions C1-C3 mainly contained xylanase activities (C1, 60 U). Details of the enzyme composition of the different complexes will be presented

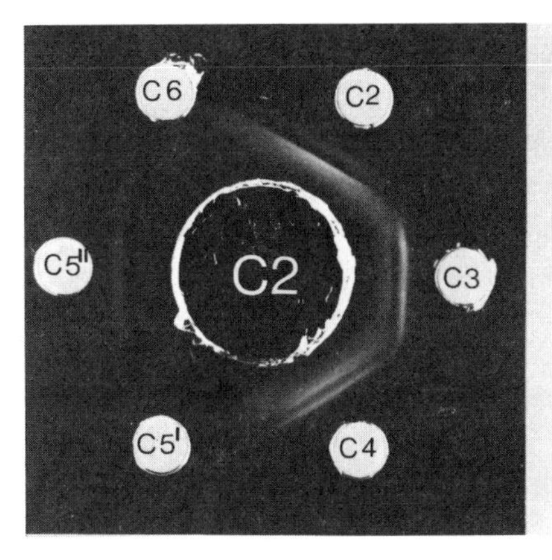


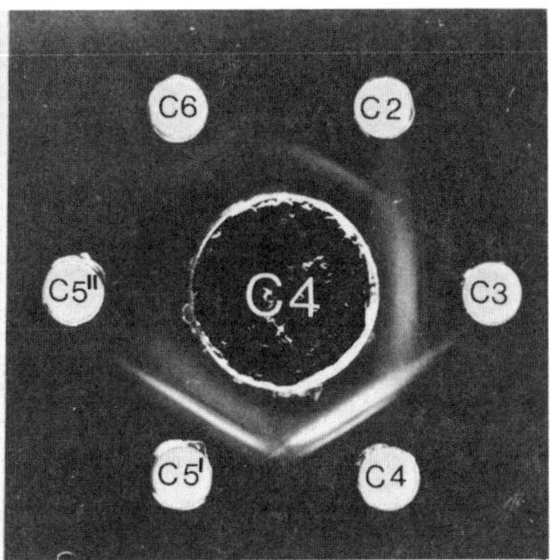


Fig. 5: Immunodiffusion of antibodies to C2 (center well) against C2, C3, C4, C5', C5", C6 exoproteins (outer wells).

Fig. 6: Immunodiffusion of antibodies to C4 (center well) against C2, C3, C4, C5', C5", C6 exoproteins (outer wells).

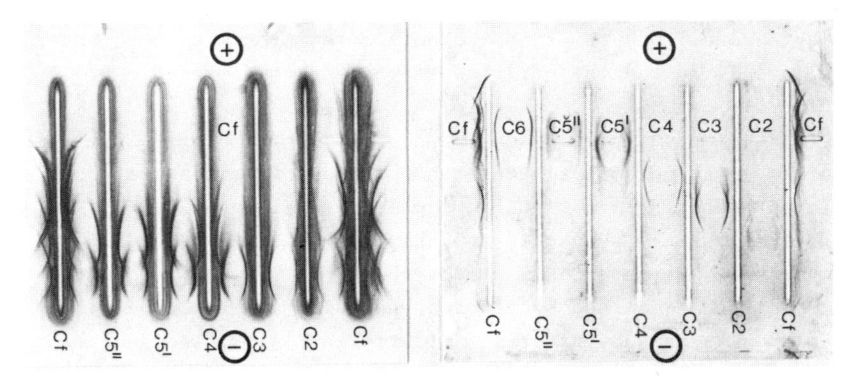


Fig. 7: Immunoelectrophoresis of culture fluid of T. reesei (cathodic direction). Antibodies to total exoproteins and to C2-C6 are indicated.

Fig. 8: Immunoelectrophoresis of purified protein complexes C2-C6 (cathodic direction) and culture fluid (outer lanes). Antibodies to corresponding protein complexes separated are indicated.

elsewhere. Summing up, homogeneous fractions obtained by preparative PAGE can contain different enzyme activities, e.g. fraction C2; xylanase, β-glucosidase, CMCase).

Serological results obtained by double diffusion techniques substantiate the view of the presence of different exoproteins within homogeneous fractions rather than that of single proteins. Antibodies raised against C2 gave precipitation lines with protein complexes C1-C5 (Fig. 5). Similar results were obtained with antibodies raised to C4. Common antigenic determinants were found to be present in C4-C6 (Fig. 6). Thus it can be concluded that common antigenic determinants might be present within most of the complexes C1-C6 purified by PAGE.

Heterospecifity of antibodies raised against purified fractions C1-C6 was documented by immunoelectrophoresis (Fig. 5, 6). Culture fluids were separated by cathodic runs with β-alanine-KOH, pH 4.3, as separation system. Antibodies raised to homogeneous fractions C1-C6 were filled sidewise into antibody application slots. Precipitation lines were stained with Coomassie blue. They showed polyspecific reaction with both the culture fluid and the purified exoprotein complexes C2-C5" (Fig. 5-8).

Heterogenity in composition of homogeneous fractions was documented when fraction C1-C6 were preparatively prepared by PAGE and subsequently subjected to analytical IEF. Macroheterogenity covering a pH-range 3-8 was observed (Fig. 9). Position of β-glucosidase localized by fluorescence with 4-methylumbelliferyl-β-glucoside as substrate is indicated in Fig. 9. In conclusion, results for 7 homogeneous protein fractions obtained by preparative PAGE in gradient gels (5-25 %) using an acidic separation system in the cathodic direction give evidence that homogenity in the case of exoproteins in T. reesei seems to be an expression for purified enzyme complexes rather than for individual enzymes.

3.2 Characteristics of exoprotein complexes preparatively purified from anodic separations with PAGE (Tris-HCl system, pH 8.8)

When separated with a alkaline discontinuous buffer system six protein complexes were obtained using the blotting technique with microcrystalline cellulose sheets. Fig. 3 gives the Coomassie-stained blot. Fig. 4 shows four purified complexes. Characteristics in short are given in Table 3. All fractions A1-A6 degraded cellulose (Avicel, F.P., CMC and cotton) and xylane though varying in relations and profiles of sugars degraded. Table 3 qualitatively lists soluble sugars formed after 1 h of enzymatic hydrolysis detected by sugar chromatography. Fig. 10 shows the spectrum of sugars formed upon Avicel, CMC and xylane hydrolysis by fraction A1. Care should be taken interpreting the CMCase test on the basis of reducing sugars only. Using a DA-X4-20 resin for sugar chromatography and washing the column with 1 M $H_3BO_3$, pH 9.45, most of the reducing sugars formed by CMC digestion were unknown in nature (Fig. 11). This result gives some servere objections against the use and the validity of the artificial source CMC substrate for an (endo)-cellulase test generally.

All A1-A6 fractions were able - as an expression of 'endo'-cellulase activity - to produce 'short fibers' or cellulose crystallites from filter paper or cotton fibers (Fig. 12). Fissure processes were found concomitantly proceeding with swelling phenomena. Fig. 13 shows a TEM micrograph of cellulose crystallites. Crystallinity of the particles produced were documented by X-ray diffraction and infrared spectroscopy. Different banding patterns of degraded cotton fibers were obtained by isopycnic Percoll gradients (unpublished results).

Heterogeneous composition of the homogeneous fractions A1-A6 prepared by gradipore PAGE was registered by immunology. Fig. 14, 15 gives the precipitation lines of antibodies to A1 and A3 after diffusion against the corresponding antigenes. Antibodies filled in the central well (about 20 µg) reacted with purified exoprotein complexes A1-A6 (3 µg) arranged in the

| Enzympräparation | aryl-ß-Glucosidase | Cellobiase | FPA | CMC-ase |
|---|---|---|---|---|
| C 2 | 0.65 | 0.078 | 0.0 | 0.21 |
| C 3 | 0.9 | 0.08 | 0.0 | 0.00 |
| C 4 | 18.1 | 4.6 | 0.02 | 0.81 |
| C 5' | 34.7 | 15.7 | 0.14 | 12.80 |
| C 5'' | 17.3 | 9.3 | 0.20 | 25.30 |
| C 6 | 0.6 | 0.57 | 0.07 | 2.80 |

Table 1: Enzyme activities (U) of fractions C2-C6 preparatively isolated by PAGE.

| | FPA | | | CMC | | |
|---|---|---|---|---|---|---|
| | Oligomere | Cellobiose | Glucose | Oligomere | Cellobiose | Glucose |
| C 2 | --- | --- | --- | --- | --- | --- |
| C 3 | --- | --- | --- | --- | --- | --- |
| C 4 | (+) | "Spur" | ++ | + | ++ | +++ |
| C 5' | --- | + | ++ | --- | --- | --- |
| C 5'' | --- | + | ++ | --- | --- | --- |
| C 6 | --- | + | + | Cello-triose | +++ | + |

Table 2: Qualitative estimate of sugars produced upon F.P. and CMCdigestion by fraction C2-C6.

outer wells.

Heterogenity in composition of purified complexes A1-A6 was demonstrated additionally by immuno-electrophoresis. Culture fluid exoprotein complexes were separated with Tris-borate buffer, pH 8.3, and reacted with antibodies to purified complexes A1-A6. The pattern formed between different lanes showed coincidences and similarties in expression supposing heterospecific composition of antibodies raised to the different homogeneous fractions each (Fig. 16). When purified complexes A1-A6 were electrophoretically separated and reacted with the corresponding antibodies immunological similarities between neighboured fractions are obvious (Fig. 17). Formation of single to four-fold precipitation lines at the same migration height give evidence for a heterogeneous composition of the antibodies formed to A1-A6 (Fig. 17).

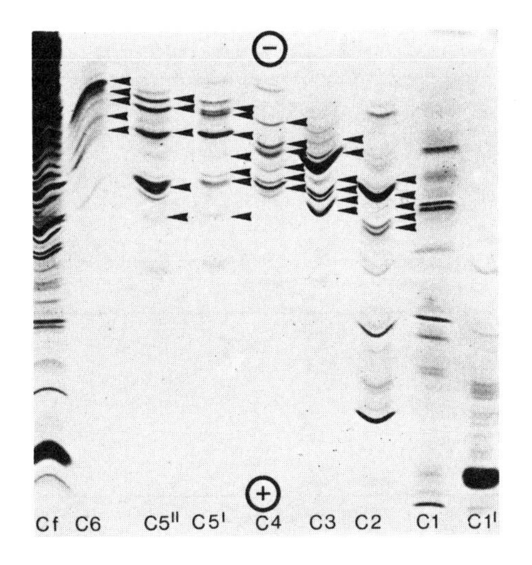


Fig. 9: Preparative IEF of fractions C1-C6 isolated by preparative PAGE. Position of β-glucosidase (positive to 4-methylumbelliferyl-β-glucoside) is indicated by arrows.

### 3.3 Characteristics of exoproteins preparatively isolated by IEF

For demonstration of the multi-enzyme complex nature of purified, homogeneous fractions, analyses were extended to preparative IEF in flat-beds using Ultrodex as supports and Servalyte as carrier ampholytes. The complete culture fluid was separated by preparative IEF using three separations (pH 2-5, 3-6 and 6-8). Gel beds were fractionated with a steel grid, yielding 30 fractions per separation. 90 fractions were tested for distribution of 7 (8) enzyme activities. Cellulolytic activities were assayed using Avicel, CMC, filter paper F.P. and cotton as substrates. Moreover, xylanase, xylosidase, β-glucosidase and acidic protease activities were included in the distribution of enzyme acitivities eluted from the serially sectioned, focused gel beds. It should be noted here that the limitation of the interpretation of this experiment is given by

| Sample | Cellooligomeres | Cellobiose | Glucose | Xylooligomeres | X2 | X1 |
|---|---|---|---|---|---|---|
| | | | Filterpaper (Whatman I) | | | |
| A 1 | - | ++ | + | - | + | + |
| A 2 | - | + | ++ | - | - | - |
| A 3 | + (G3) | +++ | (+) | - | - | - |
| A 4 | - | ++ | + | - | - | - |
| A 5 | + (G3) | ++ | (+) | - | - | - |
| A 6 | - | + | + | - | - | - |
| | | | CMC | | | |
| A 1 | + (G3-G5) | ++ | (+) | - | - | - |
| A 2 | + (G3-G4) | ++ | + | - | - | - |
| A 3 | + (G3-G4) | ++ | + | - | - | - |
| A 4 | + (G3-G6) | +++ | +++ | - | - | - |
| A 5 | + (G3-G6) | +++ | +++ | - | - | - |
| A 6 | + (G3-G6) | +++ | +++ | - | - | - |
| | | | Avicel | | | |
| A 1 | - | ++ | ++ | - | (+) | (+) |
| A 2 | - | (+) | ++ | - | - | (+) |
| A 3 | (+) G3 | ++ | + | - | - | - |
| A 4 | - | + | + | - | - | - |
| A 5 | + G3 | + | + | - | - | - |
| A 6 | - | + | + | - | - | (+) |

Table 3: Enzymatic degradation of cellulose by 6 protein complexes (A1-A6). (Qualitative pattern of sugarchromatographic separations.)

the number of enzyme tests used. It seems most likely, that testing for more enzyme activities would yield an even more precise - i.e. a more complete - result.

Table 4-6 give the distribution of enzyme activities tested over the total pI-range of Trichoderma culture fluid. The resumée derived from this experiment is in short that heterogenity - expressed in terms of presence of several enzyme activities per pI-step tested - was found generally distributed and that - as will be shown later by the titration curve experiments - it cannot be exclusively explained by the phenomenon of microheterogenity or the occurrence of isoenzymes alone.

Fractions of common pI sharing different enzyme activities are underlined (Table 4-6).

To demonstrate the purity of the fractions tested they were refocused on ultrathin gels. Part of the refocused fractions are given in Fig. 18. These fractions were tested for immunological reactions towards antibodies to pure fractions of C2, a mainly xylanase containing complex, C6 a predominantly β-glucosidase containing complex, and A2 a complex enriched in cellulase activities. Results of immunodiffusion of A2 towards the individual fractions obtained after IEF show both heterogenity and common antigenic determinants of the single antibodies used covering the more or less the complete pH-gradient as given in Fig. 19.

## 3.4 Demonstration of multi-enzyme complex composition using titration curves

Data obtained by preparative PAGE and IEF on purified homogeneous fractions document their multi-enzyme complex nature. To demonstrate the multi-enzyme character of homogeneous fractions directly the titration curve technique in presence of 6 M urea and 6 M urea-octylglucoside was assayed for a homogeneous fraction obtained after IEF. To follow the position of individual enzymes the titration curve technique was combined with a Zymogram technique: CMC or xylane were included in thin agarose films, β-glucosidase activity was observed with uv-light (350 nm) spraying the titration plates with methylumbelliferyl-β-D-glucoside. The complex we purified by preparative IEF formed one homogeneous band upon refocusing under urea- and octylglucoside-free conditions. The apparent pI of the complex was 5.7. Enzyme activities referred to the total protein content of this complex were found as follows: β-glucosidase 11,70 U (p-nitrophenol-β-D-glucoside) and 18,70 U (cellobiose as substrate). Low activities were detected against the different cellulose substrates: F.P. 0,19 U, CMC 0,37 U, cotton 0,11 U. Xylanase activity was 6,78 U. This fraction was free of any xylosidase and acidic protease activity.

Titration curves of this complex in the presence of 6 M urea

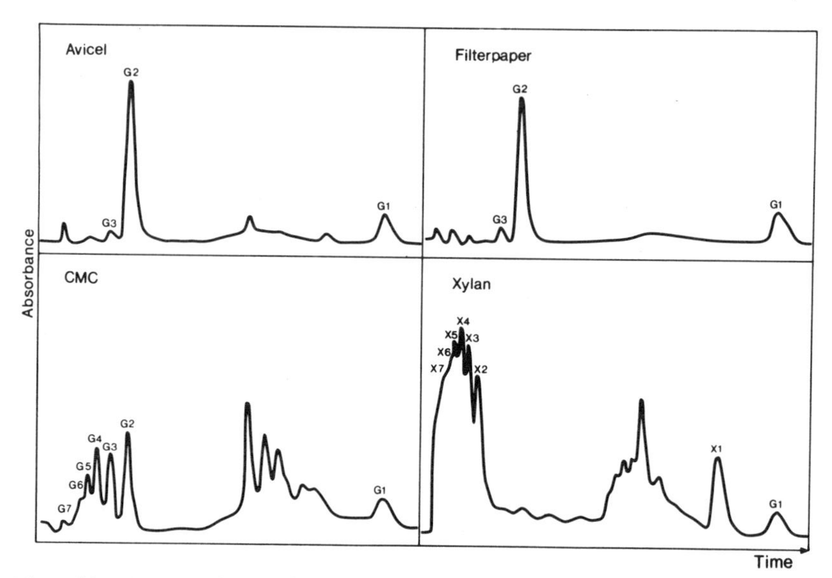


Fig. 10: Separation of soluble sugars after enzymatic degradation of different sources of cellulose and of xylane by fraction A1.

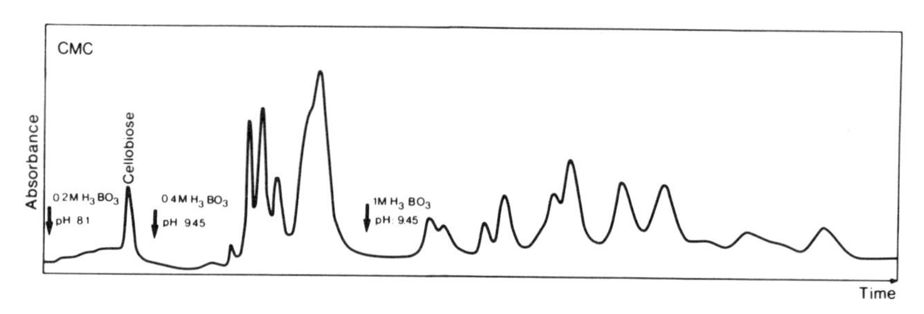


Fig. 11: Elution pattern of unknown reducing sugars obtained after a CMC digest by fraction A2.

display one main complex fraction and second faintly expressed protein fraction both sharing a common pI of 5.7 (Fig. 20). The latter fraction showed relative higher cathodic mobility characteristics at pH lower than the common apparent pI and had low electrophoretic mobility in pH ranges higher than 5.7 (Fig. 20a). In urea containing gels this small band exhibited β-glucosidase characteristics whereas the main band contained both xylanase and cellulase activities as shown in the corresponding zymograms (Fig. 20b-d). The β-glucosidase crossed the cellulase-xylanase complex at pH 3.90 (Fig. 20a). When this cellulase complex was subjected to a 6 M urea-octylglucoside treatment and separated under titration curve conditions in 6 M urea containing gels (Fig. 21) a further splitting-off of three proteins from this complex with a pI of 3.65, 5.08 and 6.01 was observed (Fig. 21a). The nature of the protein with pI 3.65 is unknown. Tested for β-glucosidase activity with 4-umbelliferyl-β-D-glucoside as substrate the titration curve of β-glucosidase was found still in the same position (pI 5.70) as after urea treatment alone (Fig. 21b), but it seemed to split into an additional fraction at pH 7.30 when poststained with Coomassie (see double arrow, Fig. 21a). Urea-octylglucoside treatment affected a shift in the route of the xylanase titration curve. The pI was found to be 6.01 as demonstrated by the lay-on zymogram technique (Fig. 20c). Urea-detergent treatment resulted in the release of one cellulase fraction with a pI of 5.08 as visualized in the corresponding zymogram (Fig. 20d).

Some conclusions based on the interprepation of titration curves from a homogeneous complex of fungal exoproteins can be derived. The complex investigated was obtained by preparative IEF. It is composed at least of six proteins, sharing a common pI of 5.7 under detergent-free conditions in the titration curves.

- Urea-octylglucoside treatment of the complex and separation in the presence of urea containing gels affects a partial splitting of the complex. The splitting of the exoprotein complex and its partial, pH-dependent release of a cellulase, xylanase and β-glucosidase, can be followed and visualized

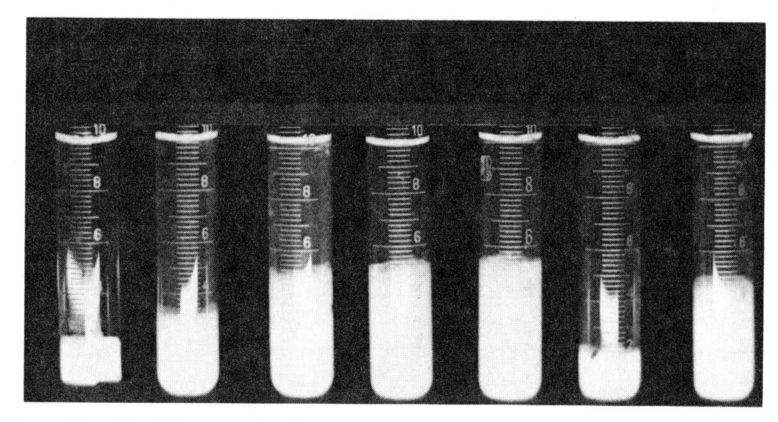

Fig. 12: Swelling and short fiber formation (control, fraction A1-A6, left to right, 15 µg protein for 15 min., Whatman I-strips, 1 x 6 cm).

Fig. 13: TEM micrograph of Pt-shadowed, cellulose crystallites formed by degradation of cotton fibers by fraction A4.

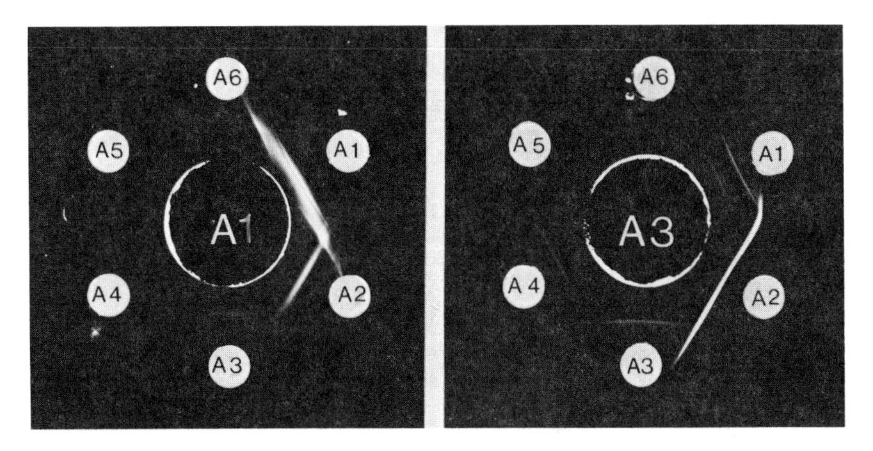


Fig. 14: Immunodiffusion of antibodies against fraction A1. Antigenes (A1-A6) are clockwise arranged in the outer wells.

Fig. 15: Immunodiffusion of antibodies against fraction A3. Antigenes (A1-A6) are clockwise arranged in the outer wells.

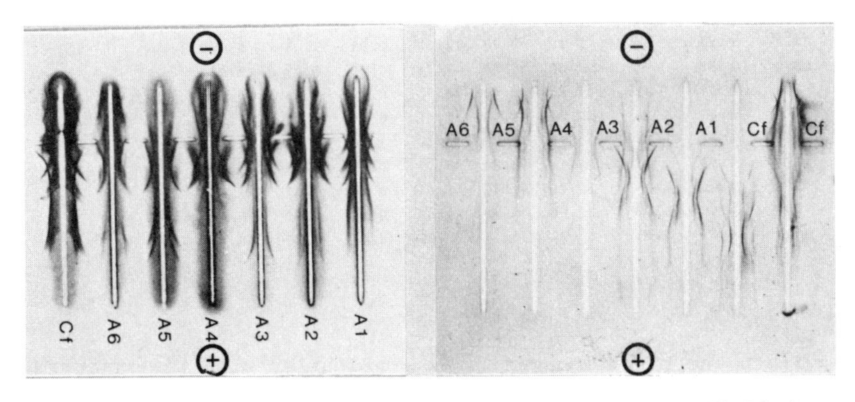


Fig. 16: Immunoelectrophoresis of T. reesei culture fluid (separated in Tris-borate, pH 8.3). Antibodies against total culture fluid (c.f.) and A1-A6 are indicated.

Fig. 17: Immunoelectrophoresis of T. reesei culture fluid (outer lanes) and A1-A6 separated in Tris-borate (pH 8.3). Antibodies against total culture fluid (c.f.) and A1-A6 are indicated.

Table 4

ISOELEKTROPHORETISCHE Trennung (1. pH-Gradient)

| Fraktion | pI | Protein ug/ml | β-gluc. | F.P. | CMC | cotton | Avicel | X'nase | X'dase | Protease |
|---|---|---|---|---|---|---|---|---|---|---|
| | | | U/mg Protein | | | | | | | |
| 1 | 2.65 | 25 | 0.00 | 2.00 | 40.54 | 0.36 | 2.34 | 6.30 | 0.000 | |
| 2 | 2.89 | 60 | 0.00 | 2.20 | 39.48 | 0.54 | 1.89 | 3.81 | 0.000 | |
| 3 | 2.99 | 165 | 0.00 | 1.00 | 15.64 | 0.31 | 1.17 | 2.77 | 0.000 | |
| 4 | 3.08 | 185 | 0.00 | 1.10 | 14.15 | 0.54 | 1.13 | 2.09 | 0.000 | |
| 5 | 3.12 | 240 | 0.00 | 1.20 | 7.31 | 0.53 | 1.18 | 2.11 | 0.000 | |
| 6 | 3.20 | 385 | 0.00 | 0.90 | 6.91 | 0.92 | 0.89 | 0.83 | 0.000 | |
| 7 | 3.28 | 210 | 0.00 | 1.90 | 12.79 | 1.46 | 1.68 | 0.98 | 0.003 | |
| 8 | 3.32 | 500 | 0.00 | 0.90 | 5.03 | 0.62 | 0.74 | 0.68 | 0.002 | |
| 9 | 3.39 | 550 | 0.00 | 0.70 | 4.31 | 0.44 | 0.56 | 1.28 | 0.010 | |
| 10 | | | | | | | | | | |
| 11 | 3.48 | 680 | 0.04 | 0.60 | 3.60 | 0.23 | 0.38 | 0.83 | 0.010 | |
| 12 | 3.57 | 760 | 0.05 | 0.30 | 3.54 | 0.18 | 0.38 | 1.04 | 0.010 | |
| 13 | 3.63 | 410 | 0.09 | 0.50 | 6.51 | 0.32 | 0.78 | 1.75 | 0.010 | |
| 14 | 3.65 | 300 | 0.11 | 0.50 | 8.92 | 0.44 | 0.93 | 3.06 | 0.010 | |
| 15 | 3.67 | 400 | 0.12 | 0.60 | 6.00 | 0.33 | 0.68 | 2.15 | 0.010 | |
| 16 | 3.76 | 240 | 0.20 | 0.70 | 10.13 | 0.53 | 1.95 | 3.93 | 0.010 | |
| 17 | 3.81 | 185 | 0.30 | 1.00 | 13.03 | 0.36 | 0.95 | 5.54 | 0.010 | |
| 18 | 3.90 | 100 | 0.90 | 1.80 | 23.03 | 0.39 | 1.59 | 9.93 | 0.020 | |
| 19 | 4.01 | 190 | 0.50 | 0.80 | 12.69 | 0.17 | 0.56 | 4.72 | 0.010 | |
| 20 | 4.04 | 140 | 0.70 | 0.70 | 17.85 | 0.33 | 0.56 | 7.13 | 0.010 | |
| 21 | 4.10 | 78 | 1.40 | 0.50 | 29.87 | 0.44 | 0.97 | 10.67 | 0.003 | |
| 22 | 4.23 | 65 | 1.40 | 0.70 | 12.53 | 0.23 | 1.01 | 8.92 | 0.000 | |
| 23 | 4.32 | 95 | 2.30 | 0.60 | 7.76 | 0.14 | 0.89 | 10.92 | - | |
| 24 | 4.39 | 33 | 5.80 | 0.60 | 26.75 | 0.26 | 1.94 | 28.73 | - | |
| 25 | 4.41 | 31 | 6.20 | 0.60 | 31.06 | 0.24 | 2.91 | 62.08 | - | |
| 26 | 4,42 | 11 | 18.40 | 0.90 | 122.30 | 1.44 | 8.11 | 61.61 | - | |
| 27 | 4.53 | 50 | 16.40 | 0.30 | 35.73 | 0.39 | 2.15 | 20.15 | - | |
| 28 | 4.60 | 125 | 12.70 | 0.40 | 18.81 | 0.34 | 0.87 | 9.44 | - | + |
| 29 | 4.68 | 230 | 22.60 | 0.30 | 9.89 | 0.13 | 0.40 | 5.51 | 0.002 | + |
| 30 | 4.87 | 780 | 55.90 | 0.13 | 0.90 | 0.03 | 0.07 | 0.46 | 0.001 | ++ |

within the titration curves in combination with the zymogram technique. The pH-dependent release of proteins observed here could explain the variance in banding patterns of β-glucosidase separated in acrylamide gels under acidic or alkaline conditions (24).

- Homogenity of fungal exoproteins after IEF and PAGE reflects purity of a complex and not of a single protein, when work-

Table 5

ISOELEKTROPHORETISCHE TRENNUNG (2. pH-Gradient)

| Fraktion | pI | Protein ug/ml | ß-gluc. | F.P. | CMC | cotton | Avicel | X'nase | X'dase | Protease |
|---|---|---|---|---|---|---|---|---|---|---|
| | | | U/mg Protein | | | | | | | |
| 1 | 3.70 | 390 | 0.04 | 0.87 | 1.15 | 0.55 | 0.65 | 3.37 | 0.01 | |
| 2 | 3.84 | 810 | 0.02 | 0.61 | 0.60 | 0.12 | 0.35 | 1.76 | 0.02 | |
| 3 | 3.99 | 1200 | 0.30 | 0.24 | 0.42 | 0.05 | 0.23 | 1.44 | 0.06 | |
| 4 | 4.12 | 800 | 0.10 | 0.65 | 0.64 | 0.14 | 0.33 | 2.28 | 0.11 | |
| 5 | 4.25 | 820 | 0.10 | 0.37 | 0.62 | 0.18 | 0.33 | 2.52 | 0.08 | + |
| 6 | 4.40 | 800 | 0.40 | 0.57 | 0.61 | 0.26 | 0.36 | 1.68 | 0.08 | + |
| 7 | 4.57 | 430 | 0.08 | 0.94 | 1.22 | 0.75 | 0.56 | 1.74 | 0.04 | + |
| 8 | 4.71 | 350 | 1.42 | 1.12 | 1.44 | 0.69 | 0.38 | 2.94 | 0.00 | + |
| 9 | 4.81 | 280 | 2.56 | 0.81 | 1.10 | 0.87 | 0.32 | 4.50 | | + |
| 10 | 4.90 | 270 | 9.16 | 0.63 | 1.63 | 0.66 | 0.46 | 3.84 | - | + |
| 11 | 5.13 | 60 | 14.30 | 1.95 | 7.13 | 1.44 | 1.61 | 5.70 | - | + |
| 12 | 5.14 | 110 | 9.61 | 0.77 | 4.21 | 0.19 | 0.37 | 8.82 | - | + |
| 13 | 5.22 | 880 | 62.03 | 0.82 | 5.70 | 0.02 | 0.48 | 7.62 | - | + |
| 14 | 5.29 | 90 | 21.50 | 0.12 | 2.84 | 0.24 | 0.30 | 10.74 | - | + |
| 15 | 5.38 | 30 | 15.69 | 0.54 | 1.85 | 0.23 | 0.71 | 27.42 | - | ++ |
| 16 | 5.42 | 30 | 7.45 | 0.58 | 1.24 | 0.30 | 0.00 | 19.08 | - | + |
| 17 | 5.45 | 30 | 2.15 | 0.58 | 1.07 | 0.14 | 0.00 | 23.94 | - | + |
| 18 | 5.52 | 130 | 8.56 | 0.16 | 0.36 | 0.08 | - | 9.30 | - | + |
| 19 | 5.57 | 160 | 11.70 | 0.19 | 0.37 | 0.11 | - | 6.78 | - | ++ |
| 20 | 5.59 | 29 | 36.46 | 0.82 | 1.88 | 0.14 | - | 28.92 | - | + |
| 21 | 5.65 | 10 | 45.09 | 0.00 | 4.31 | 0.54 | - | 55.62 | - | ++ |
| 22 | 5.71 | 16 | 3.24 | - | 1.18 | 1.11 | - | 24.72 | - | +++ |
| 23 | 5.79 | 26 | 1.00 | - | 0.59 | 0.69 | - | 17.64 | - | +++++ |
| 24 | 5.84 | 16 | 1.50 | - | 0.59 | 0.71 | - | 55.38 | - | +++++ |
| 25 | 5.93 | 27 | 1.03 | - | 0.51 | 0.39 | - | 27.72 | - | +++ |
| 26 | 6.00 | 29 | 1.54 | - | 0.51 | 0.65 | - | 44.70 | - | + |
| 27 | 6.08 | 44 | 2.31 | - | 0.61 | 0.36 | - | 32.40 | - | + |
| 28 | 6.18 | 17 | 5.68 | - | 0.56 | 0.58 | - | 49.32 | - | ++ |
| 29 | 6.35 | 62 | 1.41 | - | - | 0.63 | - | 137.36 | - | ++ |
| 30 | 6.73 | 43 | 0.70 | - | - | - | - | 105.36 | - | ++++++ |

ing under detergent-free conditions. Thus, a pI from fungal culture fluids after IEF without urea-octylglucoside treatment can be regarded as apparent pI, totaling net charges of complex involved partners (different proteins and/or glycoproteins linked to a common acidic carbohydrate and/or peptidoglycan portion).

Table 6

ISOELEKTROPHORETISCHE TRENNUNG (3. pH-Gradient)

| Fraktion | pI | Protein ug/ml | ß-gluc. | CMC | cotton | Xylanase | X'dase | Protease |
|---|---|---|---|---|---|---|---|---|
| | | | U/mg Protein | | | | | |
| 1 | 3.68 | 1800 | 0.07 | 1.54 | 0.22 | 0.67 | 0.18 | + |
| 2 | 4.39 | 750 | 0.30 | 3.97 | 0.37 | 2.10 | 0.22 | + |
| 3 | 4.48 | 660 | 1.86 | 3.99 | 0.44 | 1.65 | 0.07 | + |
| 4 | 4.69 | 154 | 13.76 | 11.95 | 0.47 | 7.31 | 0.07 | + |
| 5 | 4.99 | 124 | 17.15 | 20.66 | 0.63 | 7.62 | 0.12 | + |
| 6 | 5.16 | 152 | 12.81 | 13.06 | 0.88 | 2.76 | 0.07 | + |
| 7 | 5.70 | 84 | 25.17 | 19.96 | 2.54 | 8.85 | 0.14 | ++ |
| 8 | 5.38 | 150 | 11.70 | 9.02 | 2.32 | 0.48 | 0.06 | +++ |
| 9 | 5.53 | 106 | 1.54 | 10.57 | 3.36 | 5.97 | 0.03 | ++++ |
| 10 | 5.68 | 71 | 2.33 | 11.18 | 3.91 | 0.92 | 0.00 | +++++ |
| 11 | 5.82 | 30 | 1.77 | 26.55 | 2.54 | 18.43 | - | ++++ |
| 12 | 5.99 | 35 | 0.41 | 2.97 | 0.27 | 13.48 | - | ++++ |
| 13 | 6.25 | 33 | 0.18 | 0.00 | 0.00 | 14.14 | - | +++++ |
| 14 | 6.36 | 11 | 0.22 | - | - | 45.12 | - | +++ |
| 15 | 6.49 | 33 | 0.00 | - | - | 19.45 | - | +++++ |
| 16 | 6.64 | 88 | - | - | - | 6.52 | - | +++++++ |
| 17 | 6.78 | 23 | - | - | - | 30.11 | - | ++++++++ |
| 18 | 6.90 | 25 | - | - | - | 38.64 | - | ++ |
| 19 | 7.01 | 21 | - | - | - | 41.04 | - | + |
| 20 | 7.14 | 45 | - | - | - | 15.98 | - | + |
| 21 | 7.30 | 74 | - | - | - | 9.12 | 0.11 | + |
| 22 | 7.47 | 420 | - | - | - | 2.80 | 0.01 | + |
| 23 | 7.59 | 250 | - | - | - | 4.87 | - | - |
| 24 | 7.78 | 42 | - | - | - | 29.01 | - | - |
| 25 | 7.89 | 40 | - | - | - | 25.19 | - | - |
| 26 | 8.00 | 46 | - | - | - | 15.25 | - | - |
| 27 | 8.21 | 0.0 | - | - | - | | - | - |
| 28 | 8.23 | - | - | - | - | | - | - |
| 29 | 8.40 | - | - | - | - | | - | - |
| 30 | 8.53 | - | - | - | - | | - | - |

Evidence for the existence of a common carbohydrate portion possibly involved in linking the proteins of the complexes will be given elsewhere. The findings in part given here demonstrate that pIs published for fungal endoglucanases, exoglucanases, β-glucosidases, xylanases and xylosidases (18, 25-27) from fungal culture fluids should considered as purified complexes. Treatment of such homogeneous cellulase fractions with SDS

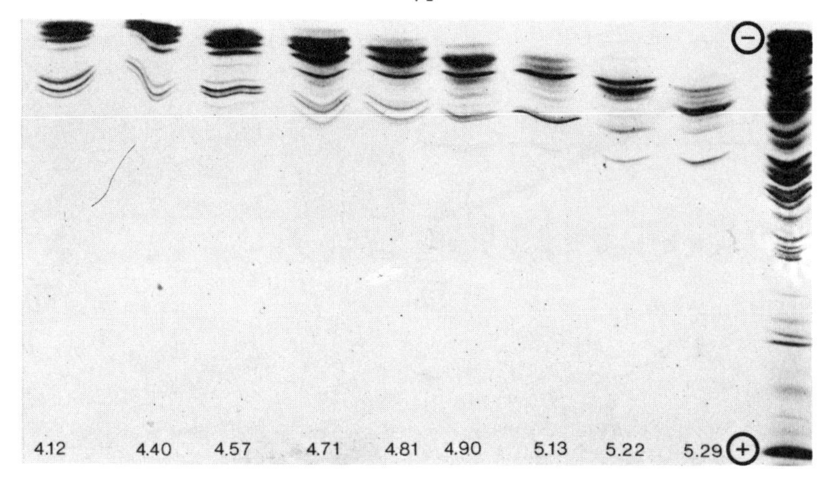


Fig. 18: IEF of part (pI 4.12 - 5.55) of the refocused series of Trichoderma culture fluid.

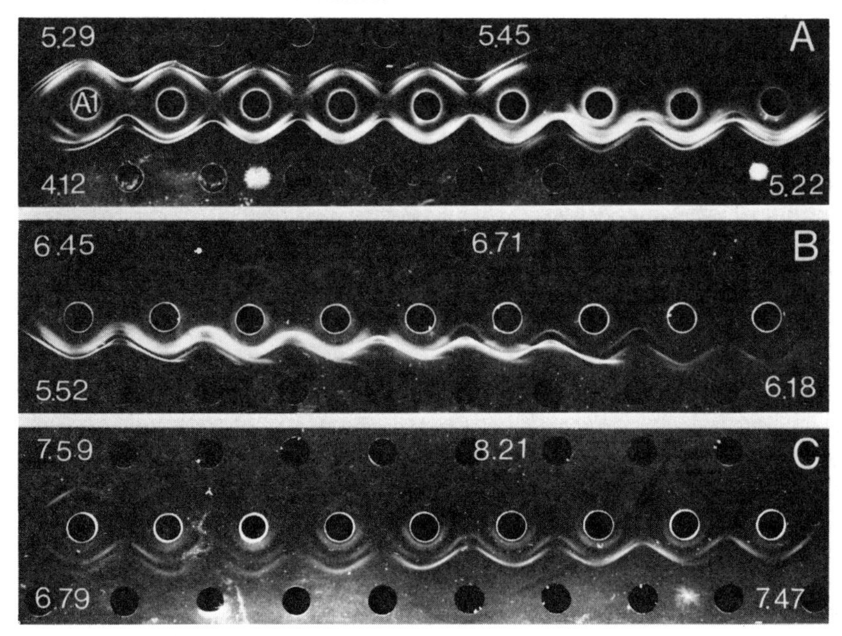


Fig. 19: Immunodiffusion of antibody against A1 towards a series (pI 3.70 - 8.53) of fractions of T. reesei culture fluid preparatively isolated by flat-bed IEF.

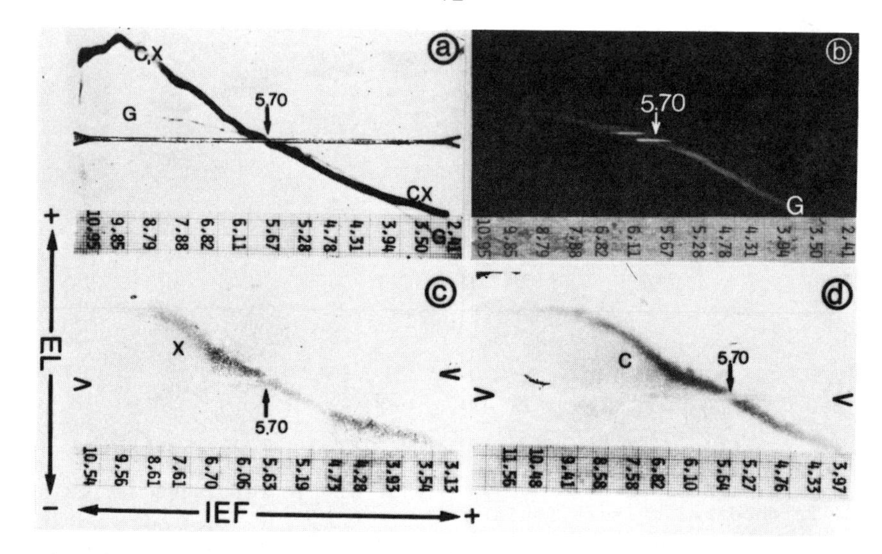


Fig. 20: Titration curve in the presence of 6 M urea of a homogeneous fraction isolated by preparative IEF. Details are described in the text.

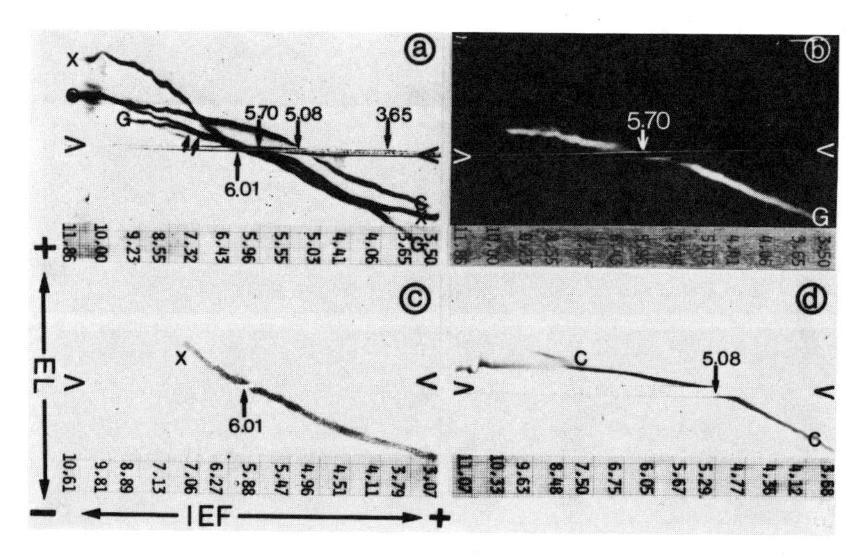


Fig. 21: Titration curve in the presence of 6 M urea of the same protein complex as given in Fig. 20 but after treatment of the sample with 6 M urea-octylglucoside.

followed by SDS-PAGE (28) does not split this complex. However, when pretreated with urea-octylglucoside it resulted - in agreement with the titration curves - in 6 proteins (results to be given elsewhere).

To degrade polymeric carbohydrate substrates it seems that the fungus favours more the release of β-1,4-glycosidic hydrolases serially aligned in complexes bound to on a common cell-wall (?) portion than the sequential excretion of single enzymes. In the light of these findings the assumed mode of cellulose degradation mediated by endo-, exocellulases and β-glucosidase (28) appears even more complicated than hitherto held, since a lectin-type binding between the cellulase complex and the sugar polymer cellulose has been bound by us. Furthermore $Ca^{++}$ influences the state of aggregation of the different complexes. Both findings will be published elsewhere.

Some aspects - though highly speculative - could favour the (multi)-enzyme complex conceipt from the 'fungalosophic' viewpoint:

- no 'waste', of excreted extracelluar proteins but a mode of recycling of multi-enzyme complexes to cell wall surfaces by means of a common proteoglycan chain and divalent cations
- a better 'survival' chance for complexes since autolytic attack of protease is reduced as protease are complex-linked
- a 'moderate' degradation of polymers
- a 'specificity' of degradation products of complex-involved enzymes (cellulases, xylanases, proteases etc.) derived from the arrangement of the enzyme in question within the complex.

Acknowledgement: The authors appreciate the valuable help of K. Irrgang for sugar chromatography.

References

1. Mandels, M.
Cellulases, A.. Rep. on Fermentation Processes, 5, p. 35 (1982)

2. Lee, Y.-H.; L.T. Fan, L.-S. Fan
Advances in Biochemical Engineering, 17, p. 131
A. Fiechter ed., Springer Verlag, Berlin (1980)

Lee, Y.-H.; L.T. Fan
Advances in Biochemical Engineering, 17, p. 101
A. Fiechter ed., Springer Verlag, Berlin (1980)

Ghose, T.K.
Biotechnol. Bioeng., 11, p. 239 (1969)

3. Eriksson, K.-E.
Microbial Polysaccharides and Polysaccharases, p. 285
R.W. Berkeley, G.W. Gooday and D.C. Ellwood eds.,
Academic Press (1979)

4. Ghose, T.K.; A.N. Pathak, V.S. Bisaria
Symposium on Enzymatic Hydrolysis of Cellulose, p. 111
M. Bailey, T.-M. Enari, M. Linko eds., SITRA,
Aulanko, Finland (1975)

5. Norman, B.E.
Microbial Polysaccharides and Polysaccharases, p. 339
R.W. Berkeley, G.W. Gooday and D.C. Ellwood eds.,
Academic Press (1979)

6. Goshe, T.
Report on Cellulase Workshop: Standard Cellulose Assay.,
Proceedings Int. Workshop held at MIT, Cambridge, U.S.A.
(1982)

7. Berghem, L.E.R.; L.G. Pettersson
Europ. J. Biochem., 37, p. 21 (1973)

8. Wood, T.M.
Biochem. J., 109, p. 217 (1968)

Wood, T.M.; S.J. McCrae
Biochem. J., 171, p. 61 (1978)

9. Halliwell, G.
Biochem. J., 79, p. 185 (1961)

10. Halliwell, G.; M. Riaz
Arch. Microbiol., 78, p. 295 (1971)

11. Sprey, B.; C. Lambert
FEMS Letters, in press

12. Ryu, Y.P.; M. Mandels
Enzyme Microbiol. Technol., 2, p. 91 (1980)

13. Furlong, C.; Ciracoglu, R.C. Willis, P.A. Santy
Anal. Biochem., 51, p. 297 (1973)

14. Reisfeld, R.A.; U.J. Lewis, D.E. Williams
Nature, 195, p. 281 (1962)

15. Radola, B.J.
Biochim. Biophys. Acta, 286, p. 181 (1974

16. Radola, B.
Electrophoresis '79 (Radola, B.J. ed.), p. 79
Walter de Gruyter 1980, Berlin - New York (1980)

17. Righetti, P.G.; E. Gianezza.
Electrophoresis '79 (Radola, B.J. ed.), p. 23
Walter de Gruyter 1980, Berlin - New York (1980

18. Berghem, L.E.R.; Pettersson, L.G.
Eur. J. Biochem., 46, p. 295 (1974)

19. Nelson, N.
J. Biol. Chem., 153, p. 375 (1944)

20. Mandels, M.; R. Andreotti, C. Roche
Biotechnol. Bioeng. Symp. 6 (E.L. Gaden, M. Mandels, E.T. Reese, L.A. Spano eds.), p. 21
John Wiley and Sons, New York (1976)

21. Saheki, T.; H. Holzer
Eur. J. Biochem., 42, p. 621 (1974)

22. Livingston, D.M.
Methods in Enzymology, 34B, p. 723
Academic Press (1974)

23. Hudson, L.; F. Hay
Practical Immunology
Blackwell Scientific Publication, Oxford (1976)

24. Shewale, J.G.
Int. J. Biochem., 14, p. 487 (1978)

25. Gorbacheva, I.V.; N.A. Rodinova
Biochim. Biophys. Acta, 484, p. 79 (1977)

26. Wood, T.M.; S.T. McCrae
Carbohydrate Res., 57, p. 117 (1977)

27. Ericksson, K.-E.; B. Pettersson
Eur. J. Biochem., 51, p. 193 (1975)

28. Reese, E.T.
Biological Transformation of Wood, p. 165
W. Liese ed., Springer Verlag, Berlin (1975)

29. Neville, D.M.
J. Biol. Chem., 246, p. 6328 (1971)

# FACTORS DETERMINING LIGNIN DECOMPOSITION AND IN VITRO DIGESTIBILITY OF WHEAT STRAW DURING SOLID STATE FERMENTATION WITH WHITHE ROT FUNGI

F. Zadražil
Institut für Bodenbiologie, Bundesforschungsanstalt für Landwirtschaft, Bundesallee 50, 3300 Braunschweig, FRG.

## Summary

Solid state fermentation of lignocellulosics with white rot fungi is a polyfactorial process with a limited number of control-possibilities. Fungal species, time, temperature of fermentation, the kind of substrate and its physical structure and chemical composition, the addition of micro- and macro-elements into the substrate, as well as the composition of the gas phase within substrate are the most important factors controlling the lignin degradation and the final in vitro digestibility of the fermented product.

The conditions for fungal growth, i.e. the physical structure of substrate (particle size, water - air - relationship in the particle and in the system, the porosity of the substrate), and the chemical composition with supplementation of nutrients are largely fixed during the preparation of the substrate. Once fermentation has started, only the compositon of the gas phase within the substrate ($O_2$ and $CO_2$) and, to a lesser extent, the water contents and the temperature of the substrate can be changed. As the thermal conductivity of lignicellulosic substrates is very low, the exchange of metabolic heat from the substrate is very difficult.

## Introduction

White rot fungi are capable of decomposing all plant cell constituents like cellulose, hemicellulose and lignin (Rypáček 1966, Kirk et al. 1980). The sequence in which these polymers are decomposed depends on the fungal species and fermentation conditions.

Selective lignin degradation was first proposed by means of upgrading lignified plant tissue for the production of animal feed (Kirk and Moore 1972, Hartley et al. 1974, Zadrazil 1977, 1980, Zadrazil et al. 1982), cellulose (paper) (Eriksson et al. 1980, Kirk and Chang 1981), and chemical feedstocks (Crawford et al. 1982).

The present paper describes the main factors determining solid state fermentation of lignocellulosics by white rot fungi with the aim of producing feed for ruminants of high digestibility.

Results

Figure 1 shows the factors determining the course of solid state fermentation of lignocellulosics. The number of possible variations between two, three or more factors is 256. However, during large-scale production the controling key-factors of fermentation are limited (Schuchardt and Zadražil 1982).

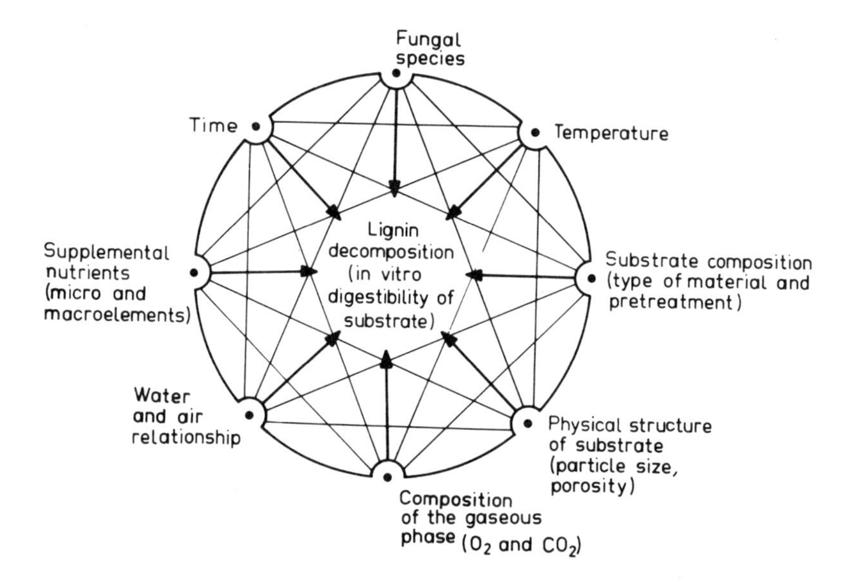


Fig. 1. Interrelation between factors determining lignin decomposition and in vitro digestibility of lignocellulosics during solid state fermentation with white rot fungi.

Fungal species: The lignin decomposition by white rot fungi depends on top of all on the fungal species and strains. Strains with high lignolytic activity have been found with the following species: Abortiborus biennis, Dichomitus squallens, Pleurotus spp., Stropharia rugosoannulata, etc.

Time: Lignin degradation is connected with secondary fungal metabolism and is induced after the colonisation of the substrate. During the primary growth mainly soluble substances, but also some cellulose and hemicellulose are utilized. The decomposition of the latter components decreases the final in vitro digestibility of the product.
It is possible to store the treated fungal substrate at a temperature of 2°C for more than 2 years without loss of the digestibility.

Temperature: The ratio of cellulose and lignin degradation in the lignocellulosics can be altered by changing the fermentation temperature. With some fungi, the temperature optimum for lignin degradation is lower than that for growth (i.e. Ganoderma applanatum - Zadražil and Brunnert 1981, Zadražil et al. 1982).

Heat-pretreatment of substrate: Heat pretreatment of the substrate eliminates part of the competitive microflora and liberates soluble organic substances from lignocellulosics. In cases where fungi of low saprophytic ability are grown, the competitive microflora reduces cellulose and hemicellulose in the substrate and thus its final in vitro digestibility.

Substrate structure: The porosity and water-air ratio of substrate particles as well as the total substrate volume correlate with the heat and gas exchange capacity. In large scale experiments these physical differences of the substrate structure are the most important factors of solid state fermentation (Zadražil and Brunnert 1982).

Addition of micro- and macro-elements: Addition of micro- and macro-elements to the lignocellulosic substrate influences the fungal growth as well as the lignin degradation and the in vitro digestibility (Reid 1979). Nitrogen, the most important element for the degradation of lignocellulosics, has a positive influence on the degradation of cellulose and hemicellulose. As lignin degradation is concerned, two groups of fungi can be differentiated, one being stimulated, the other being depressed. With 9 fungal species, addition of nitrogen caused the in vitro digestibility to decrease (Zadražil and Brunnert 1980).

$CO_2$ and $O_2$: Relatively high $CO_2$ contents in the gas phase of the substrate (15-25 vol.%) stimulate the growth of fungi decomposing wood and inhibit growth of competitive microorganisms. In the second growth phase, $O_2$ is the limiting factor for lignin degradation (Reid and Seifert 1982) and for fructification. Fructification is directly connected with intensive decomposition of all substrate components.

The scale-up of the cultivation of higher fungi on lignocellulosics is a complex problem.All parameters discussed above influence the course of solid state fermentation as well as the in vitro digestibility of the substrate at the end of the process.

References

Crawford, D.L., Barder, M.J., Pometto III, A.L. and Crawford, R.L., 1982. Chemistry of softwood lignin degradation by Streptomyces viridosporus. Arch. Microbiol., 131: 140-145.

Eriksson, K.E., Grünewald, A. and Vallander, L., 1980. Studies of the growth conditions in wood for three white-rot fungi and their cellulase-less mutans. Biotechnol. Bioeng., XXII: 363-376.

Hartley, R.D., Jones, E.C., King, N.J. and Smith, G.A., 1974. Modified wood waste and straw as potential components of animal feed. J. Sci. Fd. Agric., 25: 433-437.

Kirk, T.K. and Chang, H. 1981. Potential applications of bio-lignolytic systems. Enzyme Microb. Technol., 3: 189-196.

Kirk, T.K., Higuchi, T. and Chang, H., 1980. Lignin Biodegradation: Microbiology, Chemistry, and Potential Applications. Vol. I and II. CRC Press, Inc., Boca Raton, Florida.

Kirk, T.K., Moore, W.E., 1972. Removing lignin from wood with white-rot fungi and digestibility of resulting wood. Wood Fiber., 4: 72-79.

Reid, I.D., 1979. The influence of nutrient balance on lignin degradation by the white-rot fungus Phanerochaete chrysosporium. Can. J. Bot., 57: 2050-2058.

Reid, I.D. and Seifert, K.A., 1982. Effect of an atmosphere of oxygen on growth, respiration, and lignin degradation by white-rot fungi. Can. J. Bot., 60: 252-260.

Rypáček, V., 1966. Biologie holzzerstörender Pilze. VEB, G. Fischer, Jena.

Schuchardt. F. and Zadražil, F., 1982. Aufschluß von Lignocellu- lose durch höhere Pilze - Entwicklung eines Feststoff-Fermenters. In: Symp. Techn. Mikrobiologie, "Energie durch Biotechnologie", H. Rehm (ed.) Berlin, 422-428.

Zadražil, F., 1977. The conversion of straw into feed by Basidiomycetes. Eur. J. Appl. Microbiol., 4: 291-294.

Zadražil, F., 1980. Conversion of different plant wastes into feed by basidiomycetes. Eur. J. Appl. Microbiol. Biotechnol., 9: 243-248.

Zadražil, F. and Brunnert, H., 1980. The influence of ammonium nitrate supplementation on degradation and in vitro digestibility of straw colonized by higher fungi. Eur. J. Appl. Microbiol. Biotechnol., 9: 37-44.

Zadražil, F. and Brunnert, H. 1981. Investigation of physical parameters important for the solid state fermentation of straw by white rot fungi. Eur. J. Appl. Microbiol. Biotechnol., 11: 183-188

Zadražil, F. and Brunnert, H., 1982. Solid state fermentation of lignocellulose containing plant residues with Sporotrichum pulverulentum Nov. and Dichomitus squalens (Karst.) Reid. Eur. J. Appl. Microbiol. Biotechnol., 16: 45-51.

Zadražil, F., Grinbergs, J . and Gonzalez, A., 1982. "Palo podrido" - decomposed wood which was used as feed. Eur. J. Appl. Microbiol. Biotechnol., 15: 167-171.

# SOLID CULTURE USING ALKALI TREATED STRAW AND CELLULOLYTIC FUNGI

D.C. ULMER
Lignocellulose Group, Department of Biotechnology
Swiss Federal Institute of Technology, ETH-Hönggerberg
CH-8093 Zürich

## Summary

The possibility to enhance the protein content of wheat straw by a pretreatment to remove much of the lignin, followed by solid state fermentation (SSF) with a mixed culture of *Phanerochaete chrysosporium* (*Sporotrichum pulverulentum*) and *Candida utilis* was investigated. The expense involved in pretreating the wheat straw, the contamination problems encountered, and the low protein content of the final product indicate that SSF of wheat straw with *P. chrysosporium* is not a viable alternative to conventional protein feedstock for ruminants.

## 1. INTRODUCTION

Fermentation of lignocellulosic wastes by cellulolytic fungi for creation of single cell protein (SCP) has been proposed as a means to upgrade the material for use as an animal feed (1, 2, 3). Prior to fermentation, the material usually must be treated with hot alkali to remove the lignin and enhance the availability of the cellulose to the fungus. The pretreated material is then used as a 1% substrate in submerged culture for SCP production (2, 3, 4). The use of a low solid content and water removal from the final product makes the production of SCP from cellulosic wastes uneconomical in comparison to more conventional protein sources (5). Solid substrate fermentations (SSF) are advantageous to submerged fermentations in having lower power requirements, requiring minimum process control, having fewer contamination problems, and allowing the utilization of low-pressure air during aeration of cultures (6). This report gives a brief description of the results of trying to simplify the pretreatment and scaling up the fermentation process as a prototype for on farm application.

## 2. PRETREATMENT SIMPLIFICATION

For lab scale experiments, straw was milled to pass a 2 mm screen and then treated at 120° C for 15 to 60 min as a 5% slurry in 1% (w/v) NaOH. The treated straw was then filtered, washed, resuspended, and the suspension adjusted to the pH value desired and again filtered. This treatment not only requires a large amount of water, but also removes much of the hemicellulose which could serve as a readily available carbon source for subsequent fermentation.

To simplify this process, wheat straw was brought to 30% (w/w) dry matter using 2-10 g NaOH per 100 g straw and treated at 120° C for 20 min. This process is similar to that used for treating straw to enhance its digestibility (7). After treatment, the pH of the straw was adjusted to 4.5-5.0 using $H_2SO_4$. Attempts to grow the fungus Chaetomium cellulolyticum on straw treated in this manner were unsuccessful. Inhibition of growth could possibly be attributed to the relatively high salt concentration, inadequate neutralization of the inner portions of the straw particles, or production and accumulation of toxic compounds during the pretreatment step. A slurry process using 5% straw (w/v) in 0.5% (w/v) NaOH was used as a pretreatment in subsequent studies.

## 3. FERMENTATION PROCESS

To obtain a sufficient amount of material for rat feeding trials, trays 30 x 50 x 6 cm, which could hold 300-400 g treated straw (dry weight), were used as fermentation vessels. These trays could be stacked on top of each other so that they required relatively little space and the process was envisioned as being similar to that employed for mushroom cultivation.

A mixed culture of Phanerochaete chrysosporium and Candida utilis was used as inoculum since they had been reported to yield a higher final protein content than Chrysosporium thermophile or Chaetomium cellulolyticum (8). The amount of protein obtained during the first two trials was only 3-4%. After this, an Aspergillus infection was continually obtained from the surrounding environment. To avoid this contamination problem, we used plastic bags in which the treated straw could be placed, autoclaved and then inoculated. Results using this procedure were very erratic. Sometimes good fungal growth was obtained within 10 days, while at other times bags could be left for 30 days with no observable fungal growth, even after reinoculation. Protein content of the final product was again relatively low, 6-7%.

## 4. FUTURE OF SSF

The concept of using SSF of lignocellulose waste for production of SCP does not appear to be a realistic one. Based on material balance alone, one can see that the amount of theoretical protein attainable by SSF is relatively low. Starting with 100 g treated straw containing 18% (w/w) hemicellulose and 60% (w/w) cellulose, and assuming that the fungus could use 100% and 50% of the hemicellulose and cellulose, respectively, with a yield of 50% and the biomass containing 40% protein, this amounts to 9.6 g protein. Assuming there is a 40% loss in dry matter (not compatible with the yield value), the final product could only contain a maximum of 16% protein (9.6/60). Chahal et al. (9) reported only 65% utilization of NaOH-treated straw by C. cellulolyticum when used at a concentration of 1% (w/v) in a slurry fermentation. The yield was 0.44 and the final product contained 20% crude protein. In actuality, the yield of biomass from a SSF is certainly less than that given here. The 6-7% protein measured for our samples appears to be a realistic amount and could possibly be increased to 10-12% using optimized conditions.

The problem of an Aspergillus infection precludes the concept that SSF of wheat straw can be regarded as simple technology. Either aseptic conditions must be maintained or another fungus that can quickly establish itself and outgrow the Aspergillus must be used. The need for a slurry pretreatment and aseptic conditions, coupled with the low protein yields for SSF, makes the process an unattractive alternative to submerged cultivation or conventional protein feeds.

## REFERENCES

1. DUNLAP, C.E. (1975). Production of single-cell protein from insoluble agricultural wastes by mesophiles. In S.R. Tannenbaum and I.C. Wang, eds. Single-Cell Protein II. The MIT Press, Cambridge, MA. pp. 244-262.
2. ERIKSSON, K.E. and LARSSON, K. (1975). Fermentation of waste mechanical fibers from a newsprint mill by the rot fungus Sporotrichum pulverulentum. Biotechnol. Bioeng. 17:327-348.
3. MOO-YOUNG, M., CHAHAL, D.S. and VLACH, D. (1978). Single cell protein from various chemically pretreated wood substrates using Chaetomium cellulolyticum. Biotechnol. Bioeng. 20:107-118.
4. CHAHAL, D.S., MOO-YOUNG, M. and VLACH, D. (1981). Effect of physical and physicochemical pretreatments of wood for SCP production with Chaetomium cellulolyticum, Biotechnol. Bioeng. 23:2417-2420.
5. SENEZ, J.C., RAIMBAULT, M. and DESCHAMPS, F. (1980). Protein enrichment of starch substrates for animal feeds by solid-state fermentation. World Animal Review 35:36-39.
6. BAILEY, J.E. and OLLIS, D.F. (1977). Biochemical Engineering Fundamentals. McGraw-Hill, Inc., New York, NY, p. 163.
7. WILSON, P.N. and BRIGSTOCKE, T. (1977). The commercial straw process. Proc. Biochem. 12:17-20.
8. LUGINBUEHL, M. (1981) SCP-Produktion in Festkultur auf vorbehandeltem Stroh. Schlussbericht der Periode 1.10.1980 - 31.3.1981. COST 83/84.
9. CHAHAL, D.S., SWAN, J.E. and MOO-YOUNG, M. (1977). Protein and cellulase production by Chaetomium cellulolyticum grown on wheat straw. Dev. Ind. Microbiol. 18:433-442.

# STUDIES ON THE EXTRACELLULAR CELLULOLYTIC ENZYME SYSTEM OF CHAETOMIUM CELLULOLYTICUM

P. Fähnrich and K. Irrgang
Institut für Biotechnologie der Kernforschungsanlage Jülich

Summary

Several components of endoglucanase, ß-glucosidase and xylanase of different molecular weights were found in culture filtrate of C. cellulolyticum after fractionation by gel filtration. Besides monomeric sugars, oligosaccharides were also detected in hydrolyzates of carboxymethylcellulose and xylan with purified endoglucanase and xylanase, respectively. ß-Glucosidases hydrolyzed both aryl glucoside and cellobiose, and also exhibited transferase activity.

## 1. Introduction

The cellulolytic enzymes of C. cellulolyticum are induced in the presence of cellulose and consist of three principal types of activities, namely endo-ß-glucanase, cellulase (filter paper cellulase), and ß-glucosidase. Culture fluids also contain xylanase, ß-xylosidase and a cellobiose dehydrogenase that catalyzes the oxidation of cellobiose to cellobionic acid with simultaneous reduction of quinone (1,2). When saccharification of cellulose was carried out with crude cellulase at pH 5 and 50°C, the components of the resulting carbohydrate mixture were found to be mainly glucose, followed by cellobionic acid and cellobiose (2). In long term saccharification at 50°C especially the ß-glucosidase became inactivated. Glucose not only decreased the cellulose-solubilizing activity of crude cellulase, but also the formation of glucose from cellulose (3). To obtain a confident characterization of the cellulase complex of C. cellulolyticum it is necessary to purify the individual enzymes. This report deals with enzyme purification, molecular weight estimation, and identification of hydrolysis products of purified enzymes.

## 2. Fractionation of crude cellulase by polyacrylamide gel electrophoresis (PAGE) and gel filtration chromatography

Electrophoresis of culture filtrate revealed 16 distinct cationic proteins visualized by the Coomassie blue stain in non-denaturing polyacrylamide gels. The separated fractions were assayed for enzyme activity (Fig.1). Four peaks of endoglucanase were found and two major peaks of xylanase. Two peaks of both ß-glucosidase activity and cellulase were detected, and one peak of ß-xylosidase. Only the smaller xylanase and ß-glucosidase were completely separated, whereas the remaining enzyme activities were coincident.

Extracellular protein was further chromatographed on a Bio-Gel A 0.5m column and on a calibrated column of Bio-Gel

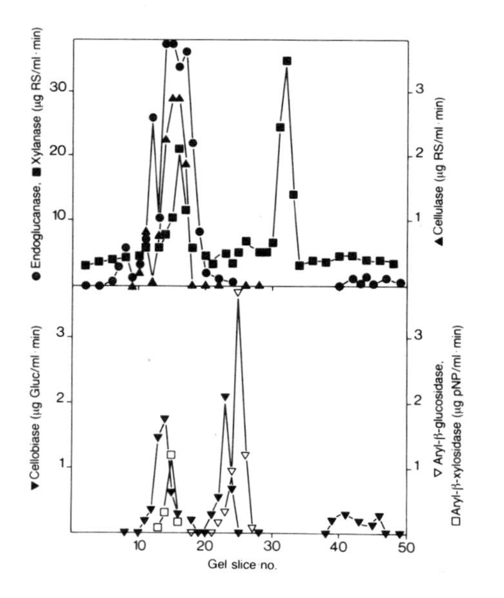


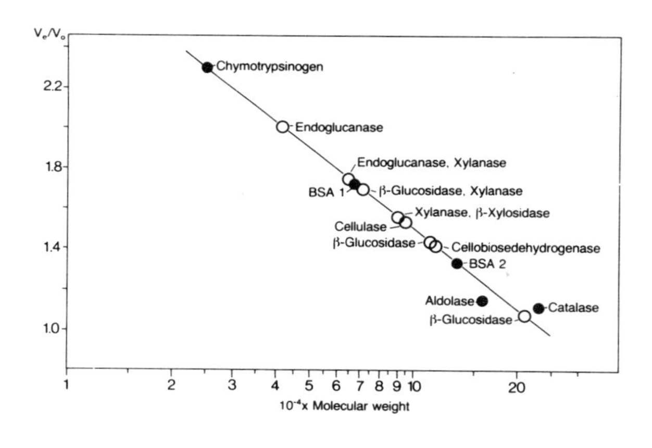


Fig.1. Activity profile of culture filtrate (200 µg protein) after PAGE. Gel system: pH 4.3–15% polyacrylamide. Gels were sliced into 1mm sections for enzyme assays (1).

Fig.2. Molecular weight estimation of enzymes. Bio-Gel P 200 column (5 x 82 cm), calibrated with chymotrypsinogen (25000), bovine albumin (BSA1 67000, BSA2 134000), aldolase (158000), catalase (232000). Buffer: 0.1 M citric acid-0.2 M $Na_2HPO_4$, 0.1 M NaCl, pH 6.8. $V_o$, void volume, determined with thyroglobulin (669000); $V_e$, eluted volume. Flow rate 5.6 ml/h; fraction size 7.6 ml.

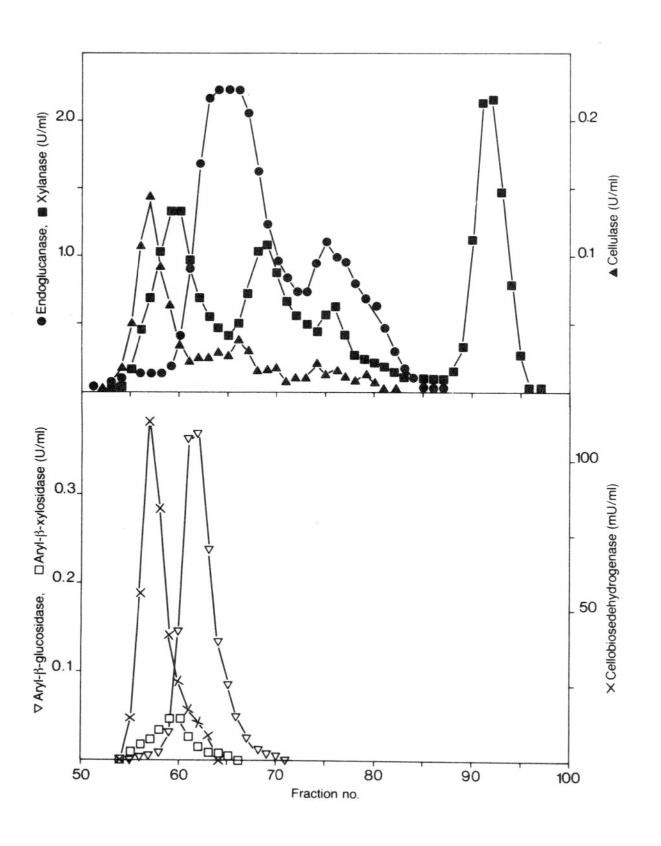


Fig.3. Distribution of enzymic activity after chromatography of a concentrated culture solution on a Bio-Gel A 0.5m column (2.5 x 90 cm). Buffer: 10 mM sodium acetate, 0.15 M NaCl, 2 mM $MgCl_2$; pH 5.8. Flow rate 15 ml/h; fraction size 9 ml.

P 200 for the estimation of apparent molecular weights of the various enzymes (Fig.2). The approximate molecular weights of two fractionated endoglucanases (E1,E2) were estimated to be about 42000 (E1) and 65000 (E2), those for three glucosidases (G1,G2,G3) were found to be (G1) 70000, (G2) 110000 and (G3) 200000. The mol.wt. of E1 compares with the value of 41000 reported for the Chaetomium thermophile endoglucanase (4). The mol.wt. of cellulase was estimated to be 95000, that for the cellobiose dehydrogenase to be 115000. Two xylanase activities were eluted from the column at the same positions as E2 and G1, respectively. A third xylanase was eluted near the cellulase, and activity towards aryl xylopyranoside was coincident with this xylanase of mol.wt. of about 90000. Fractionation on the Bio-Gel A 0.5m column resulted in the complete separation of an additional low-molecular-weight xylanase (X4), but still no complete separation of the residual enzymes was obtained (Fig.3).

## 3. Hydrolysis products of purified enzymes

Endoglucanase was also purified in a combined procedure of chromatography on concanavalin A and PAGE. The enzyme was free from glucosidase and hydrolyzed carboxymethylcellulose to glucose and cellooligosaccharides with a degree of polymerization (DP) up to 7. Products were identified by ion-exchange chromatography. Xylanase X4 likewise was further purified by PAGE, the enzyme degraded xylan to xylose and xylooligosaccharides up to DP 6. The glucosidases G1 and G4 hydrolyzed both aryl glucoside and cellobiose. Both enzymes also had transferase activity, thus tri-, penta-, and hexasaccharides were found in hydrolyzates with cellobiose.

## 4. Conclusion

The cellulase of C. cellulolyticum is a multi-component enzyme system. Endoglucanase, cellulase, glucosidase and also xylanase obviously exist as multiple enzymes. However, although protease activity was found to be low in enzyme preparations, it can not be excluded that enzyme multiplicity was due in part to proteolysis. It was not possible to separate the individual enzymes in a one-step purification procedure, such as gel filtration or PAGE. A combination of chromatographic and electrophoretic techniques could overcome this obstacle, as is shown for the purification of an endoglucanase and xylanase, respectively.

## References

1. FÄHNRICH,P. and IRRGANG,K. 1981. Biotechnol.Lett.3,201.
2. FÄHNRICH,P. and IRRGANG,K. 1982. Biotechnol.Lett.4,775.
3. FÄHNRICH,P. and IRRGANG,K. 1982. Biotechnol.Lett.4,519.
4. ERIKSEN,J. and GOKSÖYR,J. 1977. Eur.J.Biochem. 77,445.

Acknowledgement. We wish to thank Mrs. M. Gellissen for able technical assistance, and Dr. K.L. Schimz for the calibration of the Bio-Gel P 2oo column.

# PRETREATMENT AND CONVERSION OF STRAW INTO PROTEIN IN A SOLID-STATE CULTURE

W. ROSEN and K. SCHÜGERL
Institut für Technische Chemie der Universität Hannover,
Callinstr. 3, D-3000 Hannover 1

## Summary

Wheat straw was pretreated with steam at 180 °C and with $NH_3$ at room temperature and at 80 °C for varying periods of time and was ground to a maximum fiber length of 1 to 2 mm. Chaetomium cellulolyticum was cultivated on this straw in a solid-state reactor consisting of a vertical cylinder and four horizontal perforated plates aerated with humid air under nonsterile conditions. The fungus cell mass concentrations, temperature, relative humidity and pH of the culture were measured as functions of cultivation time. Cell mass concentrations achieved on straw pretreated by different methods were compared. Cellulose conversions attained in submerged and solid state reactors are about equal; however, in solid state cultures the duration of the cultivations is two to three times as long as in submerged cultures.

## 1. INTRODUCTION

The potential advantages of conducting cultivations in a solid-state system are well known: reduced fermenter volume, elimination of foaming, reduced product harvesting and drying costs. A further advantage is the opportunity to cultivate microorganisms under nonsterile conditions.

Since it was the authors' aim to develop a simple cultivation technique for farmers' use, solid-state cultures of Chaetomium cellulolyticum were investigated on pretreated wheat straw.

## 2. MATERIALS AND METHODS

The microorganism Chaetomium cellulolyticum (ATCC 32319) was obtained from the Institut für Biotechnology of KFA Jülich. The optimal growth conditions for this fungus are: 37 °C and pH 4.8 to 5.0. Pretreated wheat straw was used as a substrate. It was prepared in two different ways:

a) Steam-pretreated straw was obtained from the Institute for Wood Chemistry and Chemical Technology of Wood of the Federal Institute of Forestry and Wood Processing, Hamburg-Bergedorf. The pretreatment was carried out by steam for 5 minutes at 180 °C according to the process of this institute.

b) $NH_3$-pretreated straw was obtained in the following way: milled straw was ground to a maximum fiber length of 1 to

2 mm and was treated with 2, 4 or 8 weight % $NH_3$ at room temperature and at 60 °C for 38 to 85 days and/or for 1 year.

The straw pretreated in this way was inoculated by Chaetomium cellulolyticum and cultivated in a nonsterile solid-state reactor consisting of a vertical heat-insulated tube 70 cm in height and 16 cm in diameter, in which the cellulosic material was placed on horizontal perforated plates mixed with nutrient medium, inoculated and aerated with water-vapor-saturates air. The pH, temperature, humidity and cellulose content of the solid substance were measured as functions of cultivation time $t_F$. The cell mass content was calculated from the $N_2$ content of the protein.

## 3. RESULTS

With the steam-pretreated straw 32 weight % cell mass content was attained after 250 h at 83 to 87 % humidity and at pH 4.3 to 5.5, where the pH value was corrected by an NaOH solution.

In Table 1 the results with $NH_3$-pretreated straw are compiled. Table 1 shows that pretreatment at room temperature does not yield satisfactory results. $NH_3$-pretreatment at 60 °C is satisfactory. The increase in the duration of the pretreatment from 38 to 62 days causes only a slight increase in the protein content from 20 to 25 %. A further increase to 85 days does not improve the protein content. The increase in the $NH_3$ content from 2 % to 4 % and/or 8 % does not influence the protein content either. A comparison of submerged and solid-state cultivations indicates that 60 h are needed to attain 28 % cell mass concentration in a submerged culture with steam-pretreated straw; 120 h are needed to attain 28 % cell mass concentration in a solid-state cultivation with the same straw and 265 h to attain 26 % cell mass concentration in a solid-state cultivation with $NH_3$-pretreated straw (with 62 h pretreatment).

The authors gratefully acknowledge the financial support of the Ministry of Research and Technology of the Federal Republic of Germany, Bonn.

Table 1. Comparison of cultivations of Chaetomium cellulolyticum on wheat straw pretreated with $NH_3$ under varying conditions

| preatment | | | cultivation | | | |
|---|---|---|---|---|---|---|
| $NH_3$ (weight %) | temperature ($^{o}C$) | duration (days) | cell mass (weight %) | duration (h) | rel. humidity (%) | pH |
| 4 | 20-25 | 42 | 6.5 | 200 | 80-85 | 4.8*-5.0 |
| 4 | 20-25 | 364 | 7.8 | 240 | 83-86 | 3.9*-5.0 |
| 2 | 60 | 38 | 18.3 | 240 | 84-86 | 4.5-6.0 |
| 4 | 60 | 38 | 21.7 | 260 | 84-86 | 4.8-5.7 |
| 8 | 60 | 38 | 19.5 | 260 | 86 | 4.9-5.5 |
| 2 | 60 | 62 | 25.6 | 265 | 84-87 | 4.0*-5.5 |
| 4 | 60 | 62 | 25.6 | 265 | 86 | 4.0*-5.2 |
| 8 | 60 | 62 | 23.1 | 265 | 85-87 | 3.7*-5.4 |
| 2 | 60 | 85 | 23.3 | 260 | 85 | 4.1-5.2 |
| 4 | 60 | 85 | 26.3 | 260 | 85 | 4.3-5.6 |
| 8 | 60 | 85 | 27.8 | 260 | 85 | 4.1-5.3 |

* pH correction

# PRE-TREATMENT OF CEREAL STRAWS AND POOR QUALITY HAYS

R.D. HARTLEY and A.S. KEENE
The Grassland Research Institute, Hurley, Maidenhead, Berks, SL6 5LR

Summary

Cereal straw and ryegrass hay of low biodegradability were treated with ammonia, sulphur dioxide or sequentially with $SO_2$ followed by $NH_3$. Biodegradability, measured by a rumen liquor-pepsin or by a 'cellulase' technique, was considerably increased by all the treatments. Some of the hemicellulose and phenolic constituents of the cell walls were solubilized by the treatments, the proportion of walls degraded by cellulase being more than doubled. The final pH of straw or hay treated sequentially with $SO_2/NH_3$ could be adjusted by varying the amounts of $SO_2$ and $NH_3$.

## 1. INTRODUCTION

Our interests in this area of process development arise from the treatment by chemical methods, of poor quality forages, particularly cereal straws and poor quality hays, to increase their digestibility in the ruminant. These treatments, which have involved the use of ammonia and sulphur dioxide, lead to increases in the biodegradability of the forage cell walls measured by a commercial 'cellulase' (from *Oxyporus* spp., a Basidiomycete) having cellulase and hemicellulas activity (1). Increase in the digestibility of whole forage in the ruminant is measured *in vivo* in the animal or *in vitro* by a rumen liquor-pepsin technique (2). Although our work has been directed towards upgrading poor quality forages which can then be fed directly to the animal, the methods of treatment are likely to be useful for pre-treating graminaceous wastes to provide suitable substrates for the cultivation of microorganisms to produce high protein feeds.

## 2. RESULTS AND DISCUSSION

Barley straw (var. Sonja) and poor quality ryegrass hay (S.24) were treated with anhydrous $NH_3$ at 90°C for 16 h (3 parts of anhydrous $NH_3$ to 100 parts of dry straw or hay). The digestibility of the organic matter of the treated straw was increased by about 15 units (Table 1). Other workers have shown similar increases for cereal straws but there is little information on the treatment of poor quality hays. The content of cell walls of the barley straw decreased on $NH_3$-treatment while their degradability, determined by cellulase treatment (1), more than doubled (Table 1). The $NH_3$-treatment solubilized some of the hemicellulose constituent of the walls but the content of Klason lignin was slightly increased.

TABLE 1

Effect of treating barley straw and ryegrass hay with ammonia

| | Digestibility of the organic matter* | Cell walls (% dry matter) | Degradability of cell walls (%)† |
|---|---|---|---|
| Straw | | | |
| $NH_3$-treated | 46.5 | 79.6 | 30.3 |
| Untreated | 31.8 | 84.1 | 13.9 |
| Hay | | | |
| $NH_3$-treated | 63.8 | 72.0 | 45.9 |
| Untreated | 49.2 | 76.4 | 19.7 |

* *in vitro* rumen liquor-pepsin technique (2) † cellulase technique (1)

The nitrogen content of the straw increased from 0.6% to 1.1% after treatment. The corresponding walls had low nitrogen contents i.e. 0.1% increasing to 0.3% after treatment. The results with poor quality ryegrass hay (S.24) were similar to those of straw (Table 1), but the nitrogen contents of the whole sample and cell walls, both before and after treatment, were higher.

Earlier work showed that there was a significant correlation between the phenolic compounds released from graminaceous cell walls and the biodegradability of the modified walls (3,4). In the present work we have shown that the treatment of straw and hay with ammonia releases phenolics, including ferulic and *p*-coumaric acids, from their cell walls and that approximately 40% of the total phenolics (assessed after treatment of the walls with 1M NaOH) are released.

Further studies have shown that treatment of straw and hay with $SO_2$ also leads to large increases in biodegradability. For example, treatment of barley straw (var. Julia) with $SO_2$ (3 parts to 100 parts of dry straw) led to the digestibility of the organic matter, measured *in vitro*, being increased from 49.2 to 63.7, compared with an increase to 65.5 with $NH_3$. The pH of the $SO_2$-treated straw was 2.9 compared with 6.4 for the $NH_3$-treated straw. The effects of $SO_2$-treatment on solubilization of the hemicellulose constituent of the cell walls, and on their biodegradability measured by cellulase, were similar to those of $NH_3$-treatment. Sequential treatment with $SO_2$ followed by $NH_3$ led to larger increases in biodegradability, e.g. the digestibility of the organic matter was increased to more than 70. By varying the amounts of $SO_2$ and $NH_3$, the pH of the modified materials can be adjusted to attain suitable conditions of growth for particular microorganisms.

REFERENCES

1. HARTLEY, R.D., JONES, E.C. and FENLON, J.S. (1974). Prediction of the digestibility of forages by treatment of their cell walls with cellulolytic enzymes. J. Sci. Food Agric. 25, 947-954.

2. TILLEY, J.M.A. and TERRY, R.A. (1963). A two stage technique for the in vitro digestion of forage crops. J. Br. Grassl. Soc. 18, 104-111.

3. HARTLEY, R.D. and JONES, E.C. (1978). Phenolic components and degradability of the cell walls of the brown midrib mutant, $bm_3$, of Zea mays. J. Sci. Food Agric. 29, 777-789.

4. LAU, M.M. and VAN SOEST, P.J. (1981). Titratable groups and soluble phenolic compounds as indicators of the digestibility of chemically treated roughages. Anim. Feed Sci. Technol. 6, 123-131.

# PRODUCTION OF MYCELIAL BIOMASS ON WASTE WATER IN A ROTATING DISC FERMENTER

K. ALLERMANN[1], S. FOGH[2] and J. OLSEN[1]
[1]Institute of Plant Physiology, University of Copenhagen, Ø. Farimagsgade 2A, DK-1353 Copenhagen K., Denmark
[2]Danish Fermentation Industry Ltd., DK-2600 Glostrup, Denmark

## Summary

Growth of mycelial fungi in rotating disc fermenters were studied with the purpose of purification of industrial waste waters with difficult degradable organic compounds and production of biomass. The waste waters examined were: 1) distillery waste (cane molasses stillage) with and without pretreatment by an anaerobic fermentation, 2) lignocellulosic waste waters (black liquor from alkaline sulphite pulping of straw). The organisms used were: Sporotrichum pulverulentum, Abortiporus biennis and Aspergillus niger. The experiments have shown that the system is well suited for effluent treatment with effective retention of the biomass on the discs and BOD reductions of 62-91%.

## 1. INTRODUCTION

The disc fermenter is a bio-reactor designed with the purpose of utilizing the natural tendency of filamentous fungi to attach to and colonize solid surfaces. The use of rotating disc fermenters for the surface cultivation of mycelial fungi has been described previously (1, 2, 4).

In the disc fermenter the advantages of submerged and surface cultivations are combined. It is expected that the costs of running disc fermenters will be lower than running submerged fermentations in stirred tank reactors. This is mainly due to low energy requirements as the disc are rotated at a low speed (9 r.p.m.), and because the organism is attached (immobilized) to the discs there are no rheological problems.

The potential for using white rot fungi for purification of lignocellulosic waste waters and protein production has been investigated (6). The use of disc fermenters for effluent treatment has been suggested previously (3), and a rotating biological contactor has been used for studying colour reduction of kraft bleach effluents by fungi (5).

In this project investigations have been started with the purpose of using monocultures of filamentous fungi to treat industrial waste waters in disc fermenters; especially, waste waters with difficult degradable organic compounds. The project is supported by the Danish Council of Technology.

## 2. METHODS

Waste waters: 1) the distillery waste was a stillage from alcohol production on cane molasses, 2) the lignocellulosic waste was the black liquor from alkaline sulphite pulping of straw. Organisms: Sporotrichum pulverulentum, Abortiporus biennis, Aspergillus niger. The disc fermenter has been constructed according to (4) with a magnetic coupling instead of a shaft seal. Experiments have shown that polypropylene and polycarbonate are well suited as materials for the discs. The reported data are from fermentations at 31°C with a 2 l working volume. Rotational speed was 9 r.p.m.. The waste waters were supplemented with an inorganic salt solution and the fermenter was inoculated with a spore suspension. For some experiments cane molasses stillage was pretreated in an anaerobic sludge process without methane production.

## 3. RESULTS AND DISCUSSION

Experiments have shown that the organic load of distillery waste and lignocellulosic waste water can be reduced considerably by fermentation in a disc fermenter with S. pulverulentum and A. biennis (Table I and Figure 1B).

The distillery waste has also been treated by a two step procedure in which the raw waste was pretreated by an anaerobic fermentation. In the anaerobic stage the organic compounds were partly degraded into volatile fatty acids (VFA) which were easily metabolized by aerobic fermentation in the disc reactor. The results with A. niger are shown in Table I and Figure 1A.

The disc fermenter has been operated with thick layers of mycelium attached to the discs and with negligible free biomass in the medium; thus the biomass can be scraped of the discs and easily harvested from the medium.

The system has also been run with continuous feeding of waste water for 33 days without contamination.

Table I. Effluent treatment by mycelial fungi in a rotating disc fermenter (temp.: 31°C, pH: 5.1-5.5).

| Effluent/organism | Initial g $O_2$/l | | % reduction | | Biomass produced g/l |
|---|---|---|---|---|---|
| | COD | BOD | COD | BOD | |
| Distillery waste 1) | | | | | |
| Sporotrichum pulverulentum | 13.6 | 6.73 | 49 | 91 | 5.65 |
| Abortiporus biennis | 24.1 | 9.72 | 51 | 72 | 6.77 |
| Distillery waste 1), 2) | | | | | |
| Aspergillus niger | 10.7 | 4.97 | 70 | 85 | 3.73 |
| Lignocellulosic waste water 3) | | | | | |
| Sporotrichum pulverulentum | 9.87 | 2.40 | 27 | 62 | 0.24 |

1) Stillage from molasses.
2) After anaerobic pretreatment.
3) Black liquor from alkaline sulphite pulping of straw.

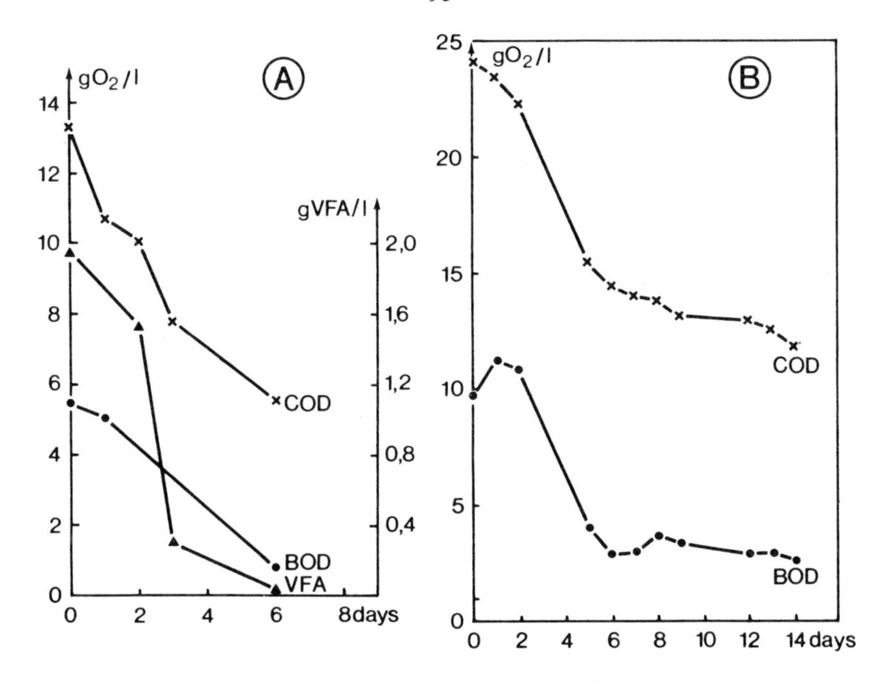


Figure 1. Treatment of distillery waste (cane molasses stillage) in a disc fermenter. A. With Aspergillus niger after anaerobic pretreatment of the waste. B. With Abortiporus biennis without pretreatment.

REFERENCES


1. ANDERSON, J.G. and BLAIN, J.A. (1980). In: Fungal Biotechnology, J.E. Smith, D.R. Berry and B. Kristiansen (Eds.) pp. 125-152. Academic Press, London.
2. ANDERSON, J.G., BLAIN, J.A., DIVERS, M. and TODD, J.R. (1980). Biotechnol. Lett. 2: 99-104.
3. ANDERSON, J.G., BLAIN, J.A., MARCHETTI, P. and TODD, J.R. (1981). Biotechnol, Lett. 3: 451-454.
4. BLAIN, J.A., ANDERSON, J.G., TODD, J.R. and DIVERS, M. (1979). Biotechnol. Lett. 1: 267-274.
5. EATON, D.C., CHANG, H.M., JOYCE, T.W., JEFFRIES, T.W., and KIRK, T.K. (1982). Tappi. 65 (6): 89-92.
6. EK, M. and ERIKSSON, K.-E. (1980). Biotechnol. Bioeng. 22: 2273-2284.

# PROTEIN ENRICHMENT OF PRETREATED LIGNOCELLULOSIC MATERIALS BY FUNGAL FERMENTATION

P. PRENDERGAST, A. BOOTH and E. COLLERAN
Department of Microbiology, University College, Galway, Ireland

Summary

Following comparative evaluation of a range of cellulolytic and lignocellulolytic fungi, protein enrichment of lignocellulosic residues was studied in submerged and solid state culture using the cellulolytic species, *Chaetomium cellulolyticum*, and the lignocellulolytic species, *Sporotrichum pulverulentum*. Although ball milling greatly enhanced wheat straw utilisation by *Chaetomium*, it had little effect on the extent of cellulose, hemicellulose and lignin utilisation by a co-culture of *Sporotrichum* and *Chaetomium*. The lignolytic ability of the fungus appeared to allow effective degradation of cellulose and hemicellulose by the co-culture and the product after 9 days culture at 37°C contained 25% crude protein D.W. NaOH pretreatment enhanced the protein enrichment of wheat straw by both species, alone or in co-culture and the product of a 9 day co-culture contained 38% protein D.W. Cellulose, hemicellulose and lignin utilisation was 75%, 91% and 52% respectively. High pressure steam pretreatment of straw, sugar cane bagasse and aspen wood allowed rapid fungal growth resulting in products containing 25-32% protein D.W. after 6 days cultivation. The carryover of soluble lignin degradation products which inhibit fungal growth on the treated substrate was noted as a disadvantage of this pretreatment method. An effective hydrogen fluoride pretreatment procedure was developed as an alternative method. The procedure allows recycling of the hydrogen fluoride and is completed within 6 minutes at room temperature. Initial experiments on the cultivation of *Chaetomium* on HF treated wheat straw revealed a rapid growth rate and the production of a product containing 27% protein D.W. after only 3 days submerged fermentation. Preliminary investigations using a solid state tumble fermenter indicate that similar growth rates may be obtained on straw substrates at 70% moisture content. Studies on the mineral nutrient requirements of the fungi indicate that all of the components of the Mandels and Reese medium (except phosphate) can be completely satisfied by the inclusion of anaerobically-digested pig slurry in the medium.

## 1. INTRODUCTION

Reliable estimates of photosynthetic activity indicate an annual world-wide productivity of over 155 billion tons of plant material(1). A considerable proportion of this material has negligible value for food, feed or industrial purposes and the consequent disposal of waste plant materials such as cereal straw, sugar cane bagasse, forestry residues, etc. by land-fill or burning represents a waste of a potentially valuable organic resource. Use of cellulosic or lignocellulosic materials for food or feed purposes also generates additional disposal problems in the handling of the resultant domestic sewage and animal slurries and manure.

A number of processing options exist for the utilisation of plant residues and food-related wastes. Two of these - anaerobic digestion to produce an alternative fuel in the form of biogas and fungal upgrading to provide a proteinaceous animal feed - have been under study in Galway for the past six years. Anaerobic digestion studies have concentrated on the development of the anaerobic filter design for rapid and efficient biomethanation of liquid agricultural wastes such as animal slurries, silage effluent, milk wastes, etc. (2,3). Laboratory and pilot scale studies have been financed by Project E of the Solar Energy Programme of the European Communities and have led to the commercialisation of the filter design and digestion process by specialist companies in Ireland and France. Anaerobic digestion and fungal protein enrichment are not mutually exclusive options. On the contrary, parallel investigations are being carried out since the biogas generated by anaerobic digestion of liquid agricultural wastes may be used to power the fungal cultivation system. Furthermore, the effluent from animal slurry digestion is rich in mineral nutrients and trace elements and may possibly be used to provide the necessary mineral addition to the fungal culture medium.

## 2. MATERIALS AND METHODS

Comparative evaluation of fungal species with respect to growth rate and substrate utilisation was carried out under submerged culture conditions. The effect of various pretreatment procedures on substrate utilisation by selected fungal species was also examined initially by submerged fermentation. Lignocellulosic and cellulosic substrates were incorporated into Mandels and Reese medium (4) at initial concentrations of 1-2% w/v. Culture volumes varied from 100 ml in shake flasks to 1.5 - 10 L in fermentation vessels. Solid state fermentation was also investigated using a 4L tumble fermenter designed and constructed in the laboratory. Innoculum volumes of 10% were routinely used and the product, after cultivation, was harvested by filtration and dried and ground to a fine powder in order to facilitate representative analysis. Crude protein was estimated by the Kjeldahl method (5) and cellulose, hemicellulose, lignin and ash content by the method of Goering and Van Soest (6). Dry matter digestibility (Tilley-Terry method) and amino acid analyses were carried out by the Irish Agricultural Institute at Dunsinea, Co. Dublin.

## 3. RESULTS AND DISCUSSION

### 3.1 Protein Enrichment of Wheat Straw

A variety of white-rot and other fungi were initially tested for their ability to utilise commercial cellulose preparations (Avicel and Solka Floc) and ball-milled wheat straw. *Chaetomium cellulolyticum* and *Sporotrichum pulverulentum* were identified as efficient cellulose and lignocellulose utilisers and all further investigations employed these two species either in mono or in co-culture. Growth of *Chaetomium* on rough-chopped (2 cm length) and ball-milled wheat straw at 37°C for 9 days indicated that ball-milling increased both cellulose and hemicellulose utilisation by *Chaetomium* and resulted in the production of a product containing 22.5% protein D.W. (Table I). Substrate utilisation and mycelial production was more efficient when *Sporotrichum* was cultivated on chopped straw and a product containing 24% protein D.W. was obtained after 9 days culture. The lignin degrading ability of this species was evidenced by a

TABLE I

Growth of Chaetomium cellulolyticum and Sporotrichum pulverulentum on wheat straw

| | % Substrate Utilisation | | | Crude Protein Production (% D.W. of 9 day pellet) |
|---|---|---|---|---|
| | Cellulose | Hemicellulose | Lignin | |
| C. cellulolyticum | | | | |
| Chopped | 40.5 | 35 | 0 | 16.5 |
| Ball-Milled | 44 | 66 | 0 | 22.5 |
| NaOH treated | 69 | 55 | 0 | 32.0 |
| S. pulverulentum | | | | |
| Chopped | 56.5 | 50 | 37 | 24 |
| Chaetomium + Sporotrichum | | | | |
| Chopped | 66.4 | 50 | 46 | 25 |
| Ball-milled | 68.4 | 65 | 46.5 | 25.5 |
| NaOH treated | 75 | 91 | 52 | 38.0 |

reduction of 37% of the initial lignin content during the growth period. Ball-milling did not appear to significantly increase the overall substrate conversion or mycelial production by Sporotrichum. Co-culture of Chaetomium and Sporotrichum on chopped and ball-milled straw yielded the results shown in Table I. It is evident from the data presented that the co-culture allowed better cellulose, lignin and hemicellulose utilisation and that ball-milling did not appreciably enhance fungal conversion of wheat straw by the co-culture. The lignolytic activity of Sporotrichum in the dual fermentation system appears to be as effective as ball-milling in enhancing the fungal upgrading of this lignocellulosic substrate. Preliminary investigation of solid state fermentation of wheat straw by the mixed fungal culture was carried out in a tumble fermenter at a moisture content of 70%. After 9 days cultivation at 37°C, the percentage utilisation of cellulose, hemicellulose and lignin was 52%, 35% and 40% respectively and a product containing 23.5% protein D.W. was obtained.

NaOH pretreatment is widely used to enhance the feed value of cereal straw for ruminants and has been used by many workers to increase substrate utilisation during fungal upgrading studies. Although NaOH greatly increases the accessibility of cellulose to enzymic attack, the treatment results in significant loss of the hemicellulose component of straw. Growth of Chaetomium in mono-culture on NaOH treated wheat straw yielded a product containing 32% protein D.W. after 9 days. As illustrated in Table I, more complete substrate utilisation was achieved by co-fermentation with Chaetomium and Sporotrichum resulting in a 9-day product containing 38% protein D.W. An initial growth rate of $0.25.h^{-1}$ was obtained. Continuation of the fermentation for a further 3 days resulted in utilisation of 95%, 92% and 57.5% of the initial hemicellulose, cellulose and lignin content respectively and yielded a final product with 43.3 protein D.W. This corresponds to a fungal biomass content in the product of 78.6%. The essential amino acid composition of the SCP product obtained by co-culture of the two fungi on chopped wheat straw is compared with the FAO standard in Table II.

Table II: Profile of Essential Amino Acids in Co-Culture Enriched Product from Wheat Straw

| | Essential Amino Acids (g per 100 g protein) | | | | | | | | |
|---|---|---|---|---|---|---|---|---|---|
| | Lys | Thr | Cyst | Meth | Val | Isol | Leu | Tyr | Phe |
| FAO Standard | 4.2 | 2.8 | 2.0 | 2.2 | 4.2 | 4.2 | 4.8 | 2.8 | 2.8 |
| Fungal Fermentation Product | 5.4 | 4.9 | N.D* | 2.5 | 6.0 | 4.9 | 4.75 | 12.6 | 6.5 |

N.D* = Not detected

3.2 Cultivation on autohydrolysed substrates

The dry matter digestibility and, consequently, the energy feed value of lignocellulosic materials is dramatically increased by the continuous high pressure steam processing system developed by Stake Technology Ltd. of Ontario (7). The process has been applied to a variety of lignocellulosic materials including straws, sugar cane bagasse, aspen wood, etc. The dry matter digestibility of wheat straw and sugar-cane bagasse is increased from an initial level of 33-35% to 60%. Utilisation of Stake processed materials as substrates for protein enrichment by *Chaetomium* was shown initially to be inhibited by the presence of soluble lignin degradation products in the autohydrolysed materials. Washing with water prior to incorporation into the fungal medium was shown to relieve the inhibition and rapid initial growth rates were obtained. Cultivation for 6 days at 37°C with *Chaetomium* in mono-culture yielded a product containing 30% protein D.W. from Stake processed straw. A product with a protein D.W. content of 26% was obtained by a similar experiment with autohydrolysed aspen wood. Treated sugar-cane bagasse was also efficiently utilised by *Chaetomium*, yielding a product of 25% protein D.W. after 6 days culture. Although good fungal conversion was obtained as a result of autohydrolysis pretreatment, the carryover of inhibitory lignin substrates presents an operational problem, particularly in the context of farm-based processes.

3.3 The effect of Hydrogen Fluoride Treatment

An investigation into other possible pretreatment processes suggested that Hydrogen fluoride, because of its effect on hydrogen bonds and glycosidic linkages (8), might prove an effective pretreatment agent. A novel process was developed which allowed efficient recycling of HF (in liquid or gaseous form) and which consequently rendered the pretreatment quite attractive from an economic standpoint. Treatment with HF at room temperature for 6 min was shown to be sufficient to increase the dry matter digestibility of barley straw from 33.4 to 62.9% and of sugar-cane bagasse from 36.1 to 63.4%. The digestibility of rice straw and sisal leaves was also increased to 59.3% and 61.4% by this process. Pretreatment with HF in the gaseous state is particularly simple and yields a fine, dry powdered product which can be utilised directly as substrate for fungal upgrading. Initial experiments on HF treated straw indicated that a product containing

27% protein D.W. could be obtained after just 3 days cultivation at 37°C.

3.4 Replacement of Mineral and Nutrient Requirements in the Fungal Culture Medium by digested slurry effluent

A series of experiments in which mineral nutrients, urea, peptone and trace elements were omitted from the Mandels and Reese medium and replaced by anaerobically digested pig slurry of equivalent nitrogen content indicated that all of the nutrients, with the exception of phosphate, could be adequately replaced by pig slurry. The phosphate requirement is partly a buffering effect and experiments are presently in hand to investigate this requirement in detail. The use of digester effluent at farm level represents a significant cost-saving and allows integration of anaerobic digestion and SCP production processes.

REFERENCES

1. BASSHAM, J.A. (1975) Cellulose Utilisation. Biotechnology Bioengineering Symposium 5, 9-20.

2. COLLERAN, E., BARRY, M., WILKIE, A. and NEWELL, P.J. (1982) Anaerobic digestion of agricultural wastes using the upflow anaerobic filter design. Process Biochemistry, 17, 12-17.
3. COLLERAN, E., BARRY, M. and WILKIE, A. (1982) The application of the anaerobic filter to biogas production from solid and liquid agricultural wastes. Energy from Biomass and Wastes VI, 443-482, I.G.T., Chicago, U.S.A.
4. MANDELS, M. and REESE, E.T. (1969) Cellulose production by *Trichoderma viride*. J. Bacteriology, 73, 269-273.
5. Standard Methods for the Examination of water and wastewater (1975), 14th Edition, American Public Health Association, Washington, D.C., U.S.A.
6. GOERING, H.K. and VAN SOEST, T. (1970) Forage fibre analysis, Agricultural Handbook No. 379, 1-16.
7. WAYMAN, M., LORA, J.H. and GULBINAS, E. (1979), Material and Energy balances in the production of ethanol from wood. ACS Symposium, Series 90, 183-201.
8. DEFAYE, J., GADELLE, A. and PEDERSEN, C. (1981), Degradation of Cellulose with hydrogen fluoride. In Energy from Biomass, 319-323. Applied Science, London.

# PROTEIN ENRICHMENT OF STARCHY MATERIALS BY SOLID STATE FERMENTATION

J.C. SENEZ
Laboratoire de Chimie Bactérienne, CN.R.S., Marseille,France.

Summary

From cassava, potato or banana residues this new process of protein enrichment by solid state fermentation gives products containing 20 % of protein and 20-25 % residual sugars. The main originality and economical interest of this simple and non-aseptic biotechnology are to be workable at the rural level in an economically integrated system combining protein enrichment with the production of the raw material and direct utilization of the product for animal breeding. Via solid state fermentation of cassava or potato, one can obtain 3 times more protein per hectare than by cultivation of soya-bean or other protein-rich pulses.

## 1. Introduction

Besides hydrocarbons and methanol, a wide variety of potential raw materials have been considered for production of single cell proteins (SCP) At this Workshop, special emphasis will be put on the potential of lignocellulosic materials, which have the advantage of being available at low cost and in quasi-unlimited quantities. But, the main limitations in this field are the lack of cellulolytic organisms with an adequate growth rate and yield and the fact that, in most instances, a costly pre-treatment is necessary.

In contrast, starchy materials, and more specifically cassava in the tropical regions or potato in temperate climates, are of obvious interest, due to both high productivity per hectare and excellent rate of bioconversion by a large variety of amylolytic bacteria, yeasts and molds. There are in the literature a considerable number of publications on that subject. However, most of the published studies have been performed only on laboratory scale, without enough consideration of economical aspects. This is namely the case of all processes involving classical fermentation in liquid medium under aseptic conditons, followed by biomass separation and drying.

As for SCP production from alkanes or methanol, the optimization of such sophisticated technology would require scaling-up to the size of industrial units having a capacity of 50 to 100 thousand tons per year, corresponding to investments well over the possibilities of most developing countries. Moreover, in the case of a large plant, the supply, transportation and storage of the raw material would pose serious problems.

## 2. The ORSTOM/IRCHA process of solid state fermentation

In view of the above considerations a quite different way of approach was adopted. A new process of protein enrichment by solid state fermentation was developed in collaboration with Drs M. Raimbault (Office de Recherche Scientifique et Technique outre-mer, ORSTOM) and F. Deschamps (Institut de Recherche en Chimie Appliquée, IRCHA).

The originality and economical interest of this process (2,3) is to be workable at the rural level, in a system combining side-by-side the production of the raw material, with protein enrichment and direct use for animal feeding. The obvious advantage of such an integrated approach is to eliminate transportation problems and to prevent intermediary profit and speculation, which would inevitably take place if either the raw material or the final product were commercialized.

All operations are performed non-aseptically in a single and simple fermenter constituted of a commercial bakery-kneader, modified for aeration and for pH and temperature control.

The first step consists in glutinization of dried starch flour by steaming at 75° C for 10 min. After cooling, water containing the nitrogen sources (urea and ammonium sulfate), mineral salts, phosphates and $2.10^7$ spores of a filamentous amylolytic fungus (Aspergillus hennebergii) is added to a total moisture content of 65 % and pH adjusted to 4.5. By gentle mechanical stirring, the inoculated substrate takes spontaneously the form of well separated granules (1 to 2 mm diameter), the mass of which is freely permeable to aeration from the bottom of the tank. The substrate is incubated at 37° C and pH 4.5 for 24-26 h under intermittent aeration and continuous adjustment of temperature , pH and water content.

The results obtained with a variety of starchy materials are shown in Table 1. In all cases the product obtained contains about 20 % protein and 20-25 % of residual sugars.

Table 1. Protein enrichment by solid state fermentation [(1)].

| | substrate | | product | |
|---|---|---|---|---|
| | protein % | carbohydrate % | protein % | carbohydrate % |
| cassava ..... | 1.0 | 90 | 18-20 | 25-30 |
| banana ..... | 6.4 | 80 | 20 | 25 |
| banana refuses | 6.5 | 72 | 17 | 33 |
| potato ...... | 5 | 90 | 20 | 35 |
| potato wastes [2] | 5 | 65 | 18 | 28 |

(1) dry flour : 100 g, water 100-120 g ; $SO_4(NH_4)_2$ : 9 g ; urea : 2.7 g $PO_4HK_2$ : 5 g.

(2) potato wastes from a fecula plant.

The overall protein/carbohydrate conversion factor is 55 % . The relatively large quantity of residual carbohydrate is due to growth limitation by the availability of extra-cellular water (1), which at the end of the culture represents only 20 % of total moisture (63 %).

Under these culture conditions, the sporulation of the mold is completely inhibited. No contamination by anaerobes takes place and the initial number of aerobic bacteria decreases by a factor of 10. This is due to excellent aerobic conditions, to acidic pH and to the fact that the disposability of free water limits the growth of bacteria well before that of the mold (1).

The amino-acide profile of the product is well balanced, with high content in lysin and thio-aminoacids. Preliminary nutritional and toxicological assays on rats and chicken proved quite satisfactory.

Similar results were obtained with a number of other amylolytic fungi

(*Asp. oryzae*, *Asp. awamori*, and various *Rhizopus* and *Penicillium* sp.) isolated from traditional fermented foods in South-East Asia, thus suggesting the possibility of utilization of protein enriched food for direct human consumption.

3. Agro-economical prospects

The agro-economical prospects are illustrated by Table 2 showing that, via solid state fermentation of cassava or potato, one can obtain about 3 times more protein per hectare than by cultivating soyabeans or other protein rich pulses.

Table 2. Comparative productivity of protein-rich feeds and protein enrichment by solid state fermentation.

| | Yield ton/ha | Protein content % | Protein ton/ha |
|---|---|---|---|
| soya-bean ............. | 1.8 | 34 | 0.6 |
| rapeseed .............. | 3.0 | 23.3 | 0.7 |
| sunflower ............. | 2.5 | 22 | 0.55 |
| horsebean ............. | 3.2 | 28 | 0.9 |
| peas ............. | 2.9 | 25 | 0.7 |
| protein enriched cassava or potato (1) ......... | 15.2 | 20 | 1.7 |

(1) yield : 40 tons/ha ; 62 % humidity, co, version factor (protein/carbohydrate : 55 %.

The ORSTOM/IRCHA process has been experimented on the large pilot scale. It is actively developed in collaboration with a French firm of bioengineering (SPEICHIM) to the size of a 700 kg (d.w.) fermenter, i.e. a capacity sufficient for breading c.a. 1400 pigs or 15,000 chicken per year.

Negotiations are in progress for rural experimentation in several tropical countries, and some of these projects will include the coupling of protein enrichment with biogas production from animal wastes, in order to make the whole system energetically independent.

It is proposed that the protein enriched fermented foods or feeds, being complementary to, but fundamentally different from the industrial SCP-s should be specifically designated by the new acronym PEFF.

REFERENCES

1. RAIMBAULT, M. (1981) Fermentation en milieu solide, ORSTOM, 291 pp.
2. RAIMBAULT, M. and GERMON, J.C. (1976) Patent B.F. N°76.06.677.
3. SENEZ, J.C., RAIMBAULT,M et DESCHAMPS,F. (1980) Protein enrichment of starchy substrates for animal feeds by solid state fermentation.World Animal Rev., F.A.O. n° 35, 36-39.

# SOLID STATE FERMENTATION OF CASSAVA WITH RHIZOPUS OLIGOSPORUS NRRL 2710

A. Ramos-Valdivia, M. de la Torre* and C. Casas-Campillo
Dept. of Biotechnology and Bioengineering, Centro de Investigacion y de Estudios Avanzados del Instituto Politecnico Nacional, Mexico.

## Summary

Cassava meal was fermented with *R. oligosporus* NRRL 2710 in order to increase its protein content. The fermentations were carried out in stirred tanks (submerged cultures) and in packed columns or trays (solid substrate cultures), under non aseptic conditions. The final product of the submerged fermentation contained 26% true protein and those of solid state fermentations contained 22.8% (packed columns) and 16.8% (trays). The specific growth rates of the fungus were lower in solid cultures than in submerged cultures, but the volumetric protein productivities of the first were higher.

## Introduction.

Protein enrichment of solid materials in solid substrate fermentation processes seems to be an option for the production of SCP (Hesseltine,1972). Among the substrates that could be utilized as raw materials, starchy substrates are especially attractive, because they are quickly metabolized by many microbial groups. A feasible substrate in tropical and subtropical countries are the cassava tubers.

The starch content of cassava tubers ranges from 64 to 80% (DM) and the protein content from 0.7 to 2.5% (DM). The plant is widely cultivated and gives a high yield (Echeverry,1975).

Among the great variety of amilolytic fungi that could be utilized, after a screening program, we selected *Rhizopus oligosporus* NRRL 2710, because it has been already utilized for human feeding, is harmless and exhibites a good protein productivity on cassava (Hesseltine and Wang, 1968; Sanchez,1981; Ramos-Valdivia,1982).

*Present address: Institut fuer Biotechnologie, Eidg. Technische Hochschule, Zuerich, Switzerland.

In this report, the effects of some important parameters on the growth of R. oligosporus NRRL 2710 in cassava solid medium under non aseptic conditions were studied. The kinetics of the submerged fermentation and the solid state fermentation were compared.

## Materials and Methods

Microorganism. *Rhizopus oligosporus* NRRL 2710 was used throughout the experiments. It was maintained by serial transfers at three monthly intervals on potatoes-dextrose agar (Bioxon).

Substrate. Four different varieties of *Manihot esculenta* Crantz were used: Mex 3, Mex 5, Mex 59 and Mex 79. These breds were developed by INIA (Instituto Nacional de Investigaciones Agricolas, Mexico). The tubers were washed, peeled, cut, cooked by steam in autoclave (100°C for 20 min),dried, grinded in a double disk attrition mill and sieved through a 4 mm sieve. The retained particles were utilized as substrate.

Culture media. From preliminary experiments, the optimal culture medium for the submerged fermentation was: $(NH_4)_2SO_4$, 1.5g; urea, 0.8g; $NaH_2PO_4$,0.3g; $MgSO_4.7H_2O$, 0.1; KCl,0.05g; $CaCl_2$,0.01g; $FeSO_4.7H_2O$,0.015g; cassava,20g (DM) and tap water 1000ml. The pH was adjusted to 3.5. The cassava solid medium was as follows: $(NH_4)_2SO_4$,7.5; urea,4g; $NaH_2PO_4$,1.5g; $MgSO_4.7H_2O$, 0.5g; KCl, 0.15g; $CaCl_2$, 0.05g, $FeSO_4.7H_2O$,0.075g; cassava 100g (DM) and tap water 100 ml.

Inocula. An inoculum of $10^6$ spores/g cassava was used throughout the experiments. The spore inoculum was prepared according to Raimbault and Alazard (1980).

Culture methods
Submerged cultures. The fermentations were carried out in a stirred tank (New Brunswick 314) with a working volume of 10l,aeration rate 0.5 VVM, agitation 300 RPM, temperature 37°C.
Solid state cultures. The solid substrate fermentations were carried out in aerated packed columns (Raimbault and Alazard, 1980) and in a tray cabinet with an aeration device.The deep of the cassava solid medium bed ranged from 12 to 15 mm.The incubation temperature was 37°C.

Analytical methods
Raw protein. Total nitrogen of the previously washed and dried samples was determined by the micro-Kjeldahl method (A.O.A.C. 47.021/70). To calculate the raw protein, the total nitrogen value was multiplied by 6.25.
True protein. The true protein content of the samples was calculated by multiplying its proteinous nitrogen by 6.25. The proteinous nitrogen was determined as follows; 25 ml distilled water and glass pearls were added to 1-2g previously

dried and milled sample. The suspension was shaked vigourously for 10 min and let rest for 30 min. Afterward 25 ml 20% Trichloroacetic acid were added. The suspension again was shaked for 10 min and let rest for 3 hours at 4°C. Subsequentially the supernatant was filtered throught filter paper Whatman 1 until a transparent filtrate was obtained. The nitrogen content of the filtrate (non proteinous nitrogen) was determined by the micro-Kjeldahl method (A.O.A.C. 47.021/70). Proteinous nitrogen was the difference between total nitrogen and non proteinous nitrogen.
Starch. It was determined according to Weinfield and Dennis (1977).
Carbohydrates. The samples were hydrolized with 6N HCl for 8 hours at 100°C. The total sugar content of the hydrolizate was determined by the phenol-sulfuric method using glucose as standard (Dubois et al.,1960).
Fat. It was determined according to A.O.A.C. 7-049/70.
Crude fiber. It was determined according to Van de Kramer and Van Ginkel's method (1952).

## Results and Discussion

Cassava varieties.
The results of the fermentation of different cassava breds with *R. oligosporus* NRRL 2710 showed that the starch content of the tubers is a fundamental factor for the protein enrichment, since the fungus is not able to use sucrose, and this sugar is the main carbohydrate of cassava tubers with a low starch percentage. For example, the fermented product of the variety Mex 69 (35.1% starch) had 18% protein while that one of Mex 79 (79% starch) had 32%. Therefore the subsequent experiments were carried out with a high starch content cassava.

Pasteurization.
Since one of the purposes of this work was to realize the fermentation under non aseptic conditions, it was tested if a pasteurization (heating at 70°C) of the cassava medium was necessary or no. The obtained results (Fig.1) indicated that for the submerged culture a pasteurization is not necessary, but it is recommendable for the solid state fermentation. This could be a result of the lower specific growth rate of the fungus cultivated in packed columns compared with that one of the submerged culture.

Moisture and aeration.
Two very important parameters for the fungal growth in solid substrate fermentations are the initial water content of the medium and the aeration rate. The water content of the media must be adequate to satisfy the water requirements of the fungi, but too much water decreases porosity and oxygen diffusion in the mass and promotes bacterial development (Raimbault and Alazard,1980). The aeration rate must provide enought oxygen to satisfy the oxygen demand of the microorganism, but it might be possible that it can not be satisfied, stil by high aeration rates, by diffusional problems.

The influence of the initial moisture of the solid medium on the growth of R. oligosporus NRRL 2710 in packed columns is shown in Fig. 2. The highest protein production and productivity corresponded to 60% initial moisture, while the growth was limited at 30% or 45% initial moisture. The influence of the aeration rate in the same system is shown in Fig. 3. It is clear that the relation between aeration rate an protein production is not a linear function.

Comparison between submerged and solid state fermentations. The time courses of the cassava fermentations in a stirred tank, in packed columns and on trays were followed. All experiments were carried out under non aseptic conditions; the solid media were heated 30 min at 70°C before inoculation and their initial water contents were adjusted to 60%. The tray fermentation was stopped when sporulation began. The fermentation kinetics are shown in Fig. 4, 5 and 6.

The true protein content of the enriched products obtained in packed columns or trays were 87.7 and 64.5% of that one of the submerged culture, while the specific growth rates of the fungus were 25.9 and 15.9% of the specific growth rate in submerged culture (Table 1).

The low specific growth rates of R. oligosporus in solid state cultures indicate that the fungus grew under limiting conditions. The main problem in solid state fermentations is diffusion, therefore it might be that oxygen was the major limiting factor. Nevertheless, the volumetric protein productivity of the solid state cultures was higher than the productivity of the sumerged one and that of the packed columns was the highest (Table 1).

The chemical analysis of the enriched cassava produced in the different fermentation devices and such of the cooked cassava are shown in Table 2. The products of the solid state fermentations had a higher carbohydrate percentage than that of submerged fermentation, therefore the carbohydrates available in solid media were not completely used up, in fact there was starch in these products.

Our results agree with those of Senez, Raimbault and Deschamps (1980), who held that it is possible to produce protein enriched products in solid state cultures, with protein contents up to 25% by using starchy materials as substrates. However, Raimbault and Alazard (1980) found that the specific growth rates of Aspergillus niger in cassava in packed columns or in submerged culture were similar, while we found that the specific growth rates of R. oligosporus were different. The reason for this discrepancy could be the different biomass ranges used to evaluate the growth rates. Raimbault and Alazard estimated the specific growth rate in the range from 0.5 to 5.84 protein, while we calculated it between 2.7 and 22.8% true protein. In packed columns it is possible,that oxygen limitation is reached, when the fungal protein concentration of the solid medium is higher than 6% (DM).

## Conclusions.

By using aerated packed columns as solid state fermentation device and R. oligosporus NRRL 2710, it is possible to increase the protein content of cassava tubers to values compatible with those of submerged cultures. The volumetric protein productivities of the solid cultures were higher than that of the submerged ones.

In spite of the fact that the growth of R. oligosporus was limited more on trays than in packed columns, the protein content of the product was acceptable.

In order to increase the protein productivity of the solid satate fermentation of cassava, further investigations are necessary. Specially those concerning to the fungal physiology in solid cultures and design of bioreactors for the scale up of the processes.

References

DUBOIS, M., GILLES, K. A. and HAMILTON, J. K. (1960). Colorimetric method for determination of sugars and related substrates. Anal Chem 28:3-6

ECHEVERRY,U.T. (1975). La yuca en la alimentacion de los colombianos. Curso de la produccion de la yuca. Instituto Colombiano Agropecuario Regional, Medellin,Colombia. 4:169-172

HESSELTINE,C.W. (1972). Solid-state fermentation. Biotechnol Bioeng 14:169-172

HESSELTINE,C.W. and WANG,H.L. (1968). Oriental fermented food made soybeans. The ninth dry bean research conference, Colorado State University, Fort Collins,Colorado,U.S.A.

HORWITZ,W.,CHICHILO,P. and REYNOLDS,H. (Eds.).(1970). Official methods of analysis of the Association of Official Analytical Chemists (A.O.A.C.).11th edition. George Barta Company, Inc. Menasha, Wiscousin, U.S.A.

RAIMBAULT,M. and D. Alazard. (1980). Culture method to study fungal growth in solid satate fermentation. European J Appl Microbiol Biotechnol 9:199-209

RAMOS-VALDIVIA,A.(1982). Enriquecimiento proteico de la yuca (Manihot esculenta crantz).M.S. thesis,Centro de Investigacion y de Estudios Avanzados del I.P.N.,Mexico

SANCHEZ,M.R. (1981). Enriquecimiento de la harina del tuberculo de yuca mediante fermentacion aerobia con hongos mucorales. Thesis.Universidad Autonoma del Edo. de Mexico, Mexico.

SENEZ,J.C.,RAIMBAULT, M. and Deschamps,F. (1980). Protein Enriched fermented feeds. VIth International Fermentation Symposium,London,Ontario,Canada.

WEINFIELD,B. and DENNIS,V.M. (1977). The determination of starch. Compost Sciences 40:27-40

VAN DE KRAMER,J.H. and VAN GINKEL,L. (1952). Rapid determination of crude fiber in cereals. Cereal Chem 29:239-251

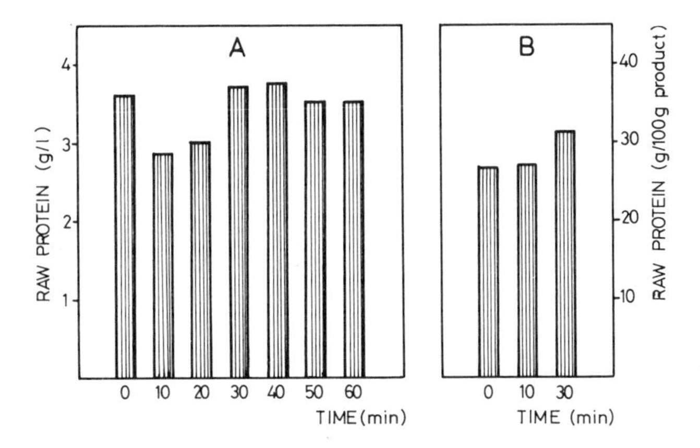


Fig. 1. Influence of the pasteurization time of the medium on the growth of R. oligosporus NRRL 2710. (A) submerged culture (B) solid substrate fermentation.

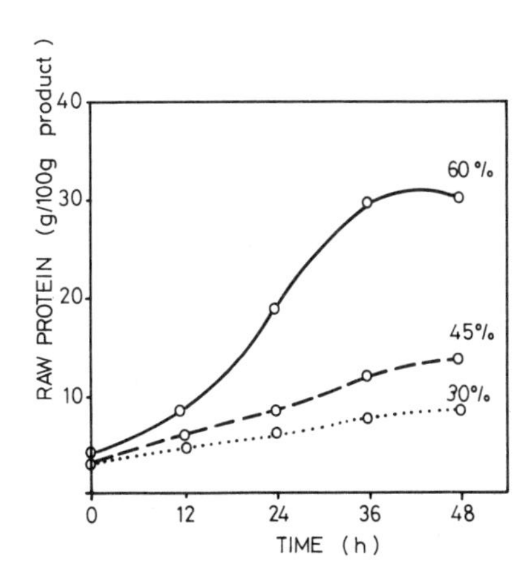


Fig. 2. Influence of the initial moisture of the cassava solid medium on the growth of R. oligosporus NRRL 2710. Parameter: initial moisture.

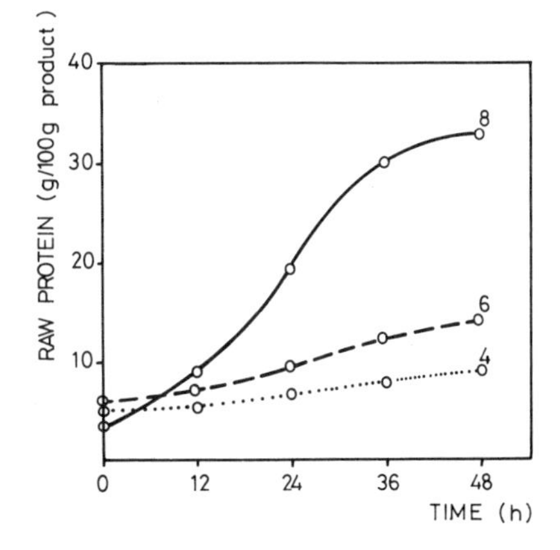


Fig. 3. Influence of the aeration rate on the growth of R. oligosporus NRRL 2710 in cassava solid medium. Parameter: aeration rate (l/h).

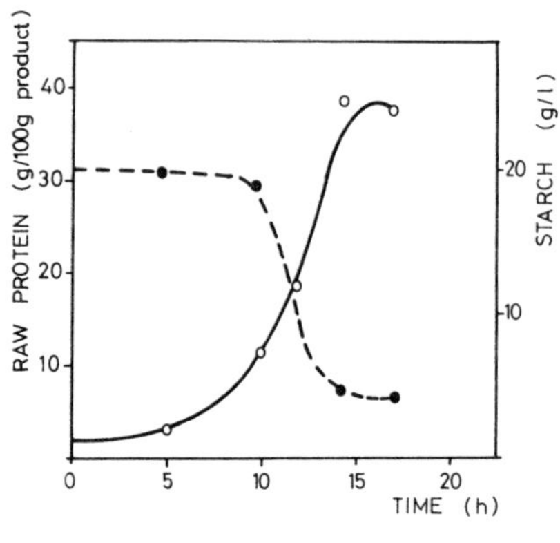


Fig. 4. Growth kinetics of R. oligosporus NRRL 2710 in cassava. Submerged fermentation. (o) protein ; (•) starch.

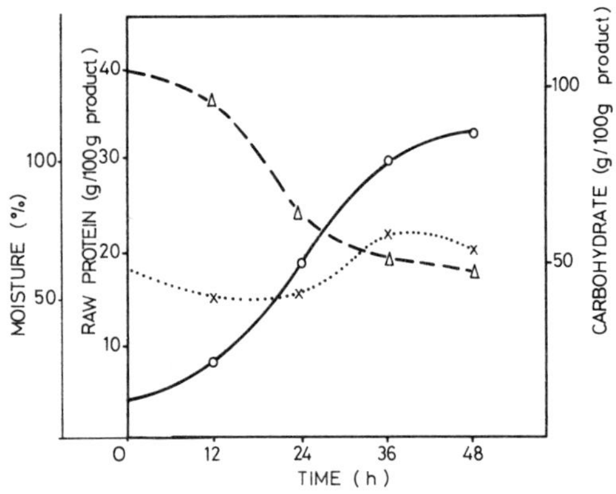


Fig. 5. Growth kinetics of R. oligosporus NRRL 2710 in cassava. Packed columns. (o) raw protein ; (Δ) carbohydrate ; (x) moisture.

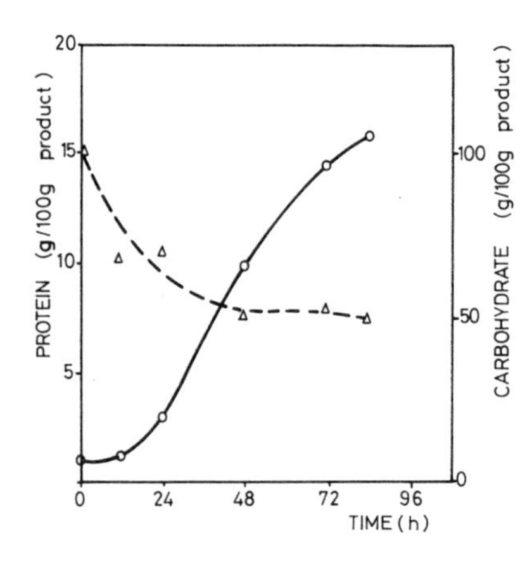


Fig. 6. Growth kinetics of R. oligosporus NRRL 2710 in cassava. Tray cabinet. (o) true protein ; (Δ) carbohydrate.

Table 1. Parameters of the submerged and solid state fermentations of cassava with R. oligosporus NRRL 2710.

| | Stirred tank | Packed columns | Trays |
|---|---|---|---|
| True protein of the end product (%) | 26.00 | 22.80 | 16.80 |
| Specific growth rate (1/h) | 0.22 | 0.06 | 0.04 |
| Total incubation time (h) | 14.00 | 48.00 | 84.00 |
| Volumetric protein productivity (g/l-h) | 0.18 | 1.20 | 0.59 |

Table 2. Chemical composition of cassava fermented with R. oligosporus NRRL 2710 and the raw material. Variety Mex 3 (18 months)

| | *Cassava | Enriched product | | |
|---|---|---|---|---|
| | %DM | Stirred tank %DM | Packed columns %DM | Trays %DM |
| Raw protein | 1.37 | 37.27 | 33.80 | 25.76 |
| True protein | 0.63 | 26.00 | 22.80 | 16.84 |
| Carbohydrates | 91.54 | 34.95 | 47.20 | 55.20 |
| Fat | 1.16 | 9.80 | 10.90 | 6.86 |
| Crude fiber | 2.94 | 7.38 | 5.36 | 6.74 |
| Ash | 2.98 | 9.22 | 1.60 | 3.81 |

*Treated as decribed in methods

# UTILISATION DE LA BAGASSE TRAITEE PAR LA SOUDE POUR LA PRODUCTION DE PROTEINES D'ORGANISMES UNICELLULAIRES

(SINGLE CELL PROTEIN PRODUCTION FROM NaOH-TREATED SUGAR CANE BAGASSE)

R.K. SEDHA, D. BERTRAND, J. DELORT-LAVAL
Institut National de la Recherche Agronomique
Laboratoire de Technologie des Aliments des Animaux
Rue de la Géraudière - 44072 NANTES - France

Résumé :

Des taux de soude variant entre 0 et 25 p. 100 de la matière sèche initiale ont été appliqués à de la bagasse, choisie comme substrat de culture de *Chaetomium cellulolyticum*. La production maximale de protéines (20 p. 100 du résidu) a été obtenue pour un taux de soude de 10 à 15 p.100, après 2 jours d'incubation et en culture submergée. L'utilisation de ce résidu, enrichi en protéines, en alimentation animale semble limitée par sa forte teneur (environ 50 p.100 de la matière sèche) en constituants pariétaux non dégradés par le microorganisme. Une étape d'extraction des protéines paraît indispensable.

## 1. INTRODUCTION

Parmi les nombreux résidus lignocellulosiques industriels ou agricoles, la bagasse (sous-produit de la canne à sucre) présente certaines caractéristiques qui pourraient en faire une matière première intéressante pour la culture de microorganismes cellulolytiques.

Cependant, son utilisation comme substrat microbien est limitée par la faible dégradabilité sous l'action des enzymes. Des essais effectués dans l'optique de l'alimentation du ruminant ont montré que la bagasse était sensible à l'action de traitements physiques (vapeur) ou chimique (traitement par la soude).

L'objet de ce travail a été d'étudier l'intérêt de la bagasse traitée par la soude comme substrat d'un microorganisme, *Chaetomium cellulolyticum* (C.c.), en vue de produire des protéines.

Les paramètres étudiés ont été les suivants : quantité de soude mise en jeu, conditions de culture du microorganisme, temps d'incubation.

## 2. MATERIELS ET METHODES

### 2.1. Traitement

La bagasse est traitée par des quantités de soude variant entre 0 et 25 p. 100 de la matière sèche (MS) initiale. 5 g de bagasse sont mis en suspension dans 95 g de solutions de soude de concentrations variées, puis chauffée à 100°C pendant 1 heure. La suspension est ensuite filtrée, lavée à l'eau distillée jusqu'à élimination de la soude libre et séchée à 80°C.

2.2. Conditions de culture

Les cultures de *Chaetomium cellulolyticum* IMI 185905 (Commonwealth Mycological Institute) sont effectuées dans des fioles de 250 ml.

Dans les essais en milieu dilué ("milieu liquide"), chaque fiole contient 50 ml de milieu de culture et 500 mg de bagasse traitée. Dans les essais en milieu concentré ("milieu solide"), ces quantités sont respectivement de 20 ml et 2,5 g. Après incubation, le contenu des fioles est filtré, lavé avec de l'eau distillée, puis séché à 80°C pour analyse.

2.3. Analyses

Les constituants pariétaux sont déterminés selon VAN SOEST et WINE, les protéines (N x 6,25) suivant la méthode de KJELDAHL et la digestibilité *in vitro* suivant TILLEY et TERRY.

Pour déterminer l'activité carboxyméthylcellulase (CMase), 1 ml du milieu filtré après incubation est incubé avec 1 ml de solution de carboxyméthylcellulose à 1 p. 100, à pH 4,8 pendant 30 minutes, à 50°C. Une unité de CMase est définie comme la quantité de sucre réducteur, exprimée en glucose, produite par 1 ml de surnageant dans les conditions de l'essai.

3. RESULTATS

3.1. Effet du traitement par la soude

Le lavage du résidu solide après traitement entraîne une disparition de matière sèche (MS) très importante (respectivement 21 et 47 p. 100 de la MS initiale pour des taux de soude de 5 et 20 p. 100).

Lorsque le taux d'application de la soude passe de 0 à 20 p. 100, la teneur en cellulose du résidu augmente de 47 à 74 p. 100. Au contraire, la teneur en hémicellulose diminue de 20 à 11 p. 100. La lignine est également fortement réduite (de 11 à 4,5 p. 100) avec une baisse particulièrement marquée pour les taux de soude compris entre 10 et 15 p. 100 (figure 1).

3.2. Développement des microorganismes

La digestibilité pour le ruminant du résidu solide après traitement de la bagasse, estimée *in vitro*, augmente sous l'action de la soude pour les taux d'application supérieurs à 5 p. 100 MS. Pour les taux inférieurs, la perte de matières organiques solubles et vraisemblablement fortement digestibles n'est pas compensée complètement par l'augmentation de la dégradabilité enzymatique de la fraction insoluble.

La courbe de la teneur en protéines de la bagasse traitée, après développement de *C.c.* en milieu liquide, en fonction du taux de soude appliqué, présente une évolution parallèle à la digestibilité *in vitro*. La teneur maximale en protéines (environ 21 p.100 MS) est atteinte pour des taux de soude de 10 à 15 p. 100 MS. La productivité maximale observée est de 1,84 g de protéines par litre de milieu. La composition du résidu pour le taux de soude de 15 p. 100 est la suivante (p. 100 MS) :cellulose : 38,3 ; hémicellulose 9,8 ; lignine : 7,2. Dans le cas du traitement par 10 p. 100 NaOH, la teneur en protéines et l'activité CMase maximales sont pratiquement atteintes après 2 jours d'incubation (figure 2).

En considérant la lignine comme marqueur non dégradé par *C.c.*, on peut estimer à 70 p. 100 la proportion de la cellulose initiale utilisée comme substrat par le microorganisme en cours de fermentation.

En milieu solide, pour une même quantité de soude mise en jeu, la teneur en protéines est en général plus faible qu'en milieu liquide, sauf pour les taux de soude de 20 et 25 p. 100.

4. CONCLUSION

Les cultures de *C.c.* sur bagasse traitée par la soude permettent d'obtenir un produit à 20 p. 100 de protéines dont l'équilibre en acides aminés semble satisfaisant. Cependant, le résidu est encore riche en constituants pariétaux indigestibles pour le monogastrique; son utilisation en alimentation animale ne pourrait être envisagée qu'après une étape d'extraction des protéines microbiennes.

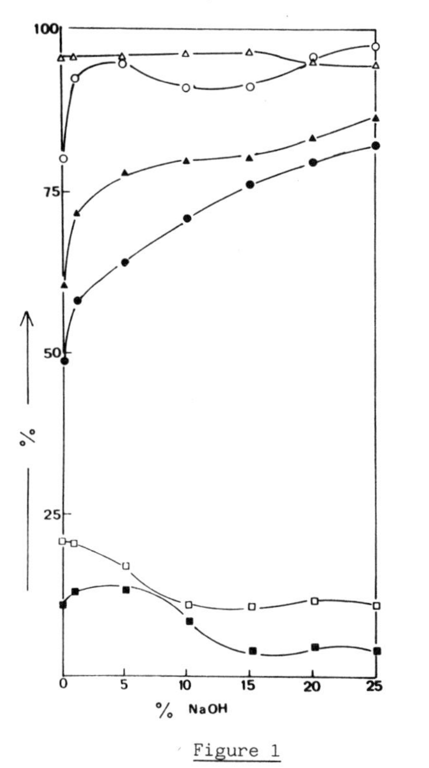


Figure 1

Effect of NaOH pre-treatment on chemical composition of bagasse. Organic matter , NDF , ADF , cellulose , hemicellulose , lignin .

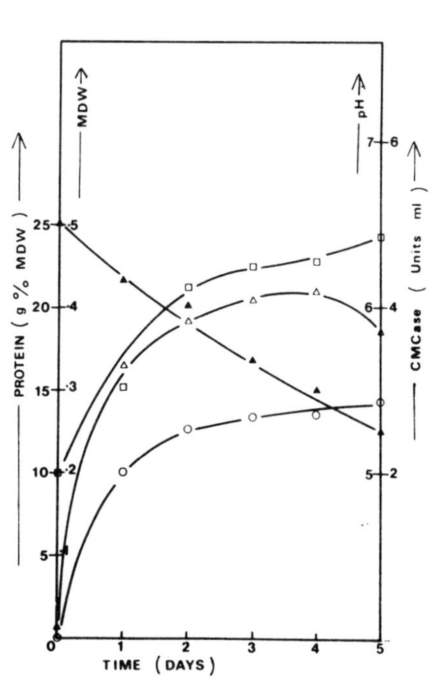


Figure 2

Fermentation time courses of C.c. with 10 % NaOH treated bagasse. Protein , CMase , pH , Mycelial dry weight (MDW) .

# CONVERSION OF AGRICULTURAL AND INDUSTRIAL WASTES FOR CELLULOSE HYDROLYSIS

M. PAQUOT, M. FOUCART, P. DESMONS, Ph. THONART
Faculté des Sciences Agronomiques de l'Etat
5800 GEMBLOUX - BELGIQUE

## Summary

Different pre-treatmentsof lignocellulosic materials have been studied in order to increase the yields of enzymatical cellulose hydrolysis. Those pre-treatments aimed at increasing cellulose accessibility in many types of agricultural or industrial wastes : straw, corn stalk, bran, draff, beet pulp.
Treatments with lime turned out to be very effective for substrates as straw or corn stalks. Moreover, beating increases the desired effect. The combination of both acid and enzymatical hydrolysis process has also been demonstrated.

## INTRODUCTION

The factors that limit enzymatical hydrolysis vary with the kind of lignocellulosic substrates. The crystallinity of cellulose and the presence of lignin are obstacles to enzymatical hydrolysis but there are other parameters : degree of polymerization, substrate hydratation, porosity ...

There are numerous pre-treatments to increase the susceptibility to enzymatic action: modification of cellulose polymorphism, alkali treatments, acid hydrolysis, mechanical treatments ...

The efficiency of the treatment varies with the kind of substrate : forestry wastes, wastepapers, agricultural and industrial wastes (straw, corn stalk, bran, draff, beet pulp, ...).

## RESULTS

### Manufacture of cellulosic pulps

These treatments aim at modifying lignin-carbohydrates interactions. They proved efficient as to hydrolysis yields for sulfite and lime treatments as well.

Nevertheless, among the substrates we have studied, only the results for straw and corn stalk are performant : about 60 % of the rough substrate could be hydrolysed, namely after a treatment with lime (corn stalks) eventually followed by milling (straw).

### Mechanical pretreatment

Beating is a mechanical treatment from papermill to increase specific surface and hydratation of fibres. Beating can be successfully applied to lignocellulosic materials such as straws treated by lime.

Modification of cellulose polymorphism

Straw hydrolysis is faster when the polymorphic form of cellulose is modified. Cadoxen method (solubilization of cellulose in a solution of cadmium oxide and ethylenediamine followed by precipitation with methanol) allows to recover and re-use solvents. 50 % of cellulose were hydrolysed within 2 hours by using cellulases from Trichoderma reesei QM 9414 (0,2 I.U.) enriched with β-glucosidase from Aspergillus niger (5 I.U.).
But some difficulties were met :

- a loss of materials (about 35 % with at least 10 % of sugars)
- a need for technical adaptation to the substrate characteristics (difficulties in solubilizing cellulose and in eliminating cadmium).

Acid pretreatment

Studies have been done with chemical pulp from hardwoods (sulfate). Acid hydrolysis is a good treatment before enzyme hydrolysis. The effects are more important for low yields of hydrolysis by acid pathway (± 10 %).

Moreover, residues coming from enzyme hydrolysis are very susceptible to acid action, which shows the complementarity of both processes.

Use of the hydrolysate

Sugar composition of the hydrolysates shows that many substrates can be valorized by fermentation. Enzymatical hydrolysis is very specific and no furfural or hydroxymethylfurfural appears in the medium. Production of single cell protein from glucose is very classical while important progresses are made in the research of microorganisms using xylose.

REFERENCES

1. HSU, T.A., LADISCH, M.R., TSAO, O.T.
Alcohol from cellulose. Chemtech, 315-319, 1980
2. THONART, Ph; PAQUOT, M; MOTTET, A.
Hydrolyse enzymatique de pâtes de papeterie. Influence des traitements mécaniques. Holzforschung 33, 197-202, 1979
3. TSHIAMALA, Z; BOUILLET, A; WATHELET, C; WATHELET, JP; PAQUOT, M;THONART, Ph. Hydrolyse enzymatique de pâtes de papeterie. Dosage des sucres obtenus en fonction des traitements mécaniques.
Holzforschung, 34, 76-79, 1980.
4. PAQUOT, M; THONART, Ph; JACQUEMIN, P; RASSEL, A.
Evolution de la rétention d'eau et de la morphologie de la fibre au cours de l'hydrolyse enzymatique de la cellulose.
Holzforschung, 35, 87-93, 1981
5. PAQUOT, M; THONART, Ph.
Hydrolyse enzymatique de la cellulose régénérée.
Holzforschung, 36, 177-181, 1982.
6. THONART, Ph; MARCOEN, JM; DESMONS, P; FOUCART, M; PAQUOT, M
Etude comparative de l'hydrolyse enzymatique et de l'hydrolyse par voie acide de la cellulose. Partie I : Morphologie du substrat en cours de l'hydrolyse enzymatique. Holzforschung, 1983 (à paraître).

Those studies were done with the supply of the EEC (R&D programme Recycling of urban and Industrial waste).

| Substrate | % cellulose + hémicellulose | Loss % | Hydrolysis % max | Hydrolysis yield |
|---|---|---|---|---|
| Straw | | | | |
| Dry milling | 66,6 | - | 0 | 0 |
| Sulfite | 85,4 | 32 | 85 | 58 |
| Lime | 80,2 | 18 | 45 | 37 |
| Lime + wet milling | 80,2 | 18 | 70 | 57 |
| Lime + beating | 80,2 | 18 | 75 | 62 |
| Bran | | | | |
| Wet milling | - | 5 | 26 | 25 |
| Lime | 90,0 | 49 | 55 | 28 |
| Sulfite | 90,6 | 72 | 65 | 18 |
| Corn Stalks | | | | |
| Wet milling | 70,1 | 5 | 13 | 12 |
| Lime | 84,8 | 31 | 91 | 63 |
| Sulfite | 93,7 | 48 | 96 | 49 |

Table 1 : Hydrolysis of straws, bran and corn stalks (pH 4,85, 48°C, 48 h)

| Substrate | HYDROLYSATE | |
|---|---|---|
| | Xylose (g/l) | Glucose (g/l) |
| Corn Stalks | | |
| Sulfite | 13,1 | 25,5 |
| Lime | 10,5 | 25,7 |
| Jute | | |
| Lime | 6,6 | 34,4 |
| Sisal - hemp | | |
| Lime | 8,7 | 34,0 |
| Straw | | |
| Sulfite | 7,0 | 17,3 |
| Lime 15' beating | 7,7 | 18,2 |
| Lime milled, 15' beating | 8,2 | 21,3 |

Table 2 : Sugar composition of different lignocellulosic materials hydrolysate.

# CELLULOSE HYDROLYSIS OF PAPERMILL SLUDGE

M. PAQUOT and L. HERMAN
Faculté des Sciences Agronomiques de l'Etat
5800 GEMBLOUX - BELGIQUE

Summary

Sludge from papermills has been considered as a potential source of sugars themselves basics materials for fermentation. A key problem in the utilisation of papermill sludge resides in the high mineral loading content (50 % of dry solids). High mineral loading hinders the acid hydrolysis route both from the point of view of the hydrolysis yield and from the fermentation efficiency.
The presence of mineral matter is less problematic for enzyme hydrolysis. But the risk of microbial infection in enzyme hydrolysis is obvious. A sterilization of the sludge seems to be necessary in this case.
An elegant way to prepare papermill sludge as a substrate for hydrolysis would also be to fractionate it on a vibrating screen, thus elimating a high proportion of the loading.

## I. INTRODUCTION

Sludge from papermills has been considered up to now being a waste product which was costly to the producer to dispose of. There are some outlets possible for which such waste has a nutritional value in the form of building materials, as a source of fibres or as a source of cellulose. This work has been done with Virginal Papermill Sludge (and supplied by Wiggins Teape S.A., Belgium).

Cellulose has been considered as a potential source of sugars, themselves basic materials for numerous fermentation routes to products such as ethanol and single cell protein.

Several different techniques were tried out on Virginal sludge to generate ethanol : acid hydrolysis followed by fermentation, enzyme hydrolysis and fermentation. The objective was throughout to typify and investigate Virginal sludge as a potential raw material for each of these modes of cellulose conversion.

## II. RESULTS

### Acid hydrolysis

A key problem in the utilisation of papermill sludge resides in its high mineral loading content. $(\frac{\text{mineral loading}}{\text{fibres}} = \pm 1)$. High minaral loading hinders the acid hydrolysis route, both from the point of view of the yield from hydrolysis and from the fermentation efficiency. The presence of mineral matter leads to a buffering effect on the sludge, so an increased acid demand is observed. Heavy metal ions are also produced in the liquid fraction of the hydrolysate.

An elegant way to prepare papermill sludge as a substrate for acid hydrolysis would be to fractionate it on a vibrating screen, thus eliminating a high proportion of the loading $(\frac{\text{mineral loading}}{\text{fibres}} = 0{,}08)$.
In this case, ethanol yield is increased three fold.

Enzyme hydrolysis

Considered as a substrate for enzyme hydrolysis, sludge exhibits some advantages and disadvantages. On the negative side, one must take note of the greater risk of microbial infection associated with this type of substrate. This may lead to a significant consumption of sugars produced during hydrolysis. Such a drawback can be overcome by first sterilising the sludge.

The presence of mineral matter remains a drawback as far as the overall yield of the hydrolysis is concerned. However, this problem is less acute than for acid hydrolysis, for the enzyme complex possesses high specificity and moreover, conditions are less severe during hydrolysis.

Hydrolysis as far as 60 % is possible with non treated sludge whereas 84 % hydrolysis is obtained when the sludge has been treated by fractionating screen. Moreover, fermentation routes does not set any problem : 94 % of the glucose is fermented without any difficulties.

REFERENCES

1. PAQUOT M., HERMANS L.
Alternatives possibles à la mise en décharge des boues de papeteries.
Un exemple : Wiggins Teape (Belgium) S.A.
Tribune du CEBEDEAU, mars 1983.

# PROTEIN ENRICHMENT OF SUGAR BEET PULP BY SOLID STATE FERMENTATION

A.DURAND ; P.ARNOUX ; O.TEILHARD de CHARDIN ; D. CHEREAU ; C.Y. BOQUIEN ; G. LARIOS DE ANDA
Station de Génie Microbiologique I.N.R.A. Dijon (FRANCE)

## Summary

The results obtained during the experiments carried out in 1982 allow to draw the following conclusions :
A - We managed to obtain by a solid-state fermentation (20 to 22% dry-matter) of about 50 hours an increase of the protein content up to 18-20% (as compared with dry-matter) in the raw beet pulps without any pre-treatment by using two nitrogen sources and a mineral solution.The mutant strain used is *Trichoderma album*.
B - The technical studies allow to define :
- a new type of pilot fermentor liable to be scaled up to production plant level. This pilot plant with a one ton pulp capacity is presently operating in Dijon.
- a simple, "rustic" enrichment process easy to incorporate into a sugar refinery.
C - The first "in vitro" nutritional assays on the products obtained show :
- a good pepsic digestibility (about 80%)
- a favourable amino-acid composition
Assays on different animals (first on lambs) are in progress.
D - At present the estimated cost price (1,15F/kg) of food enriched at 18% protein and 90% dry-matter seems to be attractive.

## 1.INTRODUCTION

The protein request in the world is continuously increasing. The deficit was about 10 millions tons in 1980 and it has been estimated at 25 millions tons in the year 2000.
France consumes 1,7 millions tons of protein annually. The French production represents only 30% of these needs. Thus it is necessary to import about 2,5 millions tons of soya cakes. The reduction of this dependency is considered in France as a political and economic priority. As a result of the soya cakes low price importated in France, the agricultural by-products cannot be economically valorized by liquid fermentation to produce proteins for feedstuffs. The difficulty comes essentially from the expenditure of energy required for the treatments before and after fermentation. For becoming competitive, it is absolutely necessary to reduce these expenditures by carrying out protein enrichment by solid state fermentation. The aim of this research was to elaborate a new prototype of pilot fermentor (one ton

capacity) liable to be extended to production plant. This pilot fermentor, presently operating in Dijon on sugar beet pulps, allows :
- to prove the technical feasability of such a process
- to perfect the process control
- to supply protein enriched pulps for nutritional assays

Two industries are collaborating on this research :
- GENERALE SUCRIERE : a sugar refinery which is interested in the industrial applications
- NORDON ET Cie : an engineering society which is interested in the construction of industrial units.

## 2.PRELIMINARY APPROACHES OF THE PROBLEM

To compete on the market of feedstuffs with soya cakes a certain number of essential choices were necessary.

### 2.1.Choice of the substrate

The choice fell on sugar beet pulps for different reasons :
- metropolitan and renewable production
- significant tonnage and geographically concentrated
- low price
- easy protein enrichment by solid state fermentation
- already present on the feedstuff market after enrichment with soya cakes.

For a sugar refinery treating 6000 tons/day of sugar beets, there is a production of about 1500 tons/day of pulps (20% dry-matter).
In France, taken as a whole, 1,1 millions tons of sugar beet pulps are available and their composition is very attractive for the purpose.

### 2.2. Choice of the microorganism

Moulds seem to be particularly well appropriated to this type of culture. Thus from a wild strain of *Trichoderma viride*, a new strain has been selected by genetic mutation : *Trichoderma album* (INRA patent). This strain is able to growth on starchy and lignocellulosic substrates. Its characteristics are : proteins content about 60% with an amino-acid composition very favourable, no toxin and unpleasant smell production, low nucleic acids content.

### 2.3. Choice of engineering

The aim of this research is to settle a reactor and a process liable to be scaled up to production plant. Thus we built in collaboration with Nordon and Co an 1 ton pilot plant in the laboratory on the model of box kiln malting system with perforated bottom, turning machine and aeration system. This pilot plant allows :
-to define and optimize an enrichment process
-to supply products for nutritional assays

## 3. RESULTS

As this research is carried out with two industries, all the details cannot be disclosed. The experiments carried out in 1982 allowed to end at the following results :

* It is technically possible to obtain a protein enrichment of sugar beet pulps by a solid-state fermentation. The raw beet pulps coming off the industry, without any pre-treatment are inoculated with mycelium of *Trichoderma album*. The inoculum is prepared in a liquid bioreactor by using cheese-whey as carbone substrate. After filtration the mycelium is mixed with raw pulps, two nitrogen sources and mineral solution.
The inoculum uptake being about 2% (1 g protein/100 g dry-matter) a protein enrichment up to 18-20% (as compared with dry-matter) is obtained in about 50 hours.
* We defined a simple,"rustic" enrichment process without sterilization easy to incorporate into a sugar-refinery. Controled variables are : temperature, pH, aeration, humidity.
* We built a new type of pilot fermentor liable to be scaled up to production plant of about 100 tons/hour. This pilot plant consists mainly of a parallelipipedic shape vessel ; three turning machines move on all the length.The pulps are supported by a perforated bottom. The air necessary for the growth of mycelium is processed in an air conditioner.
* The first"in vitro" nutritional assays on the products obtained show :

- a good pepsic digestibility (about 80%)
- a favourable amino-acid composition as compared with the traditional nitrogen sources
- a high solubility of nitrogen (about 60%)

## 4. CONCLUSIONS AND PROSPECTS

In the present state of our knowledge, an estimation of the investments and operating costs has been calculated for a plant producing about 200.000 tons/year of food enriched at 18% protein and 90% dry-matter. At present the estimated cost price (1,15 F/kg) seems to be attractive. For reducing the cost price, trials are actually carried out :

- optimization of the process
- reduction of the investment cost
- studies on preservation of pulps before and after the process of enrichment

On the nutritional aspects assays are carried on in collaboration with different zootechnical laboratories :

- April 1983 : assays on protein tanning
- October 1983 : product formulation and nutritional studies on lambs
- 1984-1985 : studies in the 1 ton pilot plant, to know how to proceed to obtain food for milking cows and monogastric animals.
- 1985 : setting on an industrial site (sugar refinery) of an experimental unit of about 10 tons.

## SUBJECT AREA 2

## SINGLE CELL PROTEIN FROM WHEY

Chairman : Z. PUHAN

Review paper :
General aspects of production of biomass by yeast fermentation from whey and permeate

Review paper :
Utilisation of whey and ultrafiltration permeates

Upgrading of mild UF-permeate by yeast fermentation - Semi-industrial trials and economy

Industrial production of S.C.P. from whey

Study of S.C.P. production from starch

Whey as a source for microorganisms / Amino acid pattern

SCP production from whey: scale-up of a process

# GENERAL ASPECTS OF PRODUCTION OF BIOMASS BY YEAST FERMENTATION FROM WHEY AND PERMEATE

O. Moebus and M. Teuber
Institut für Mikrobiologie der Bundesanstalt für Milchforschung,
D-2300 Kiel, Federal Republic of Germany

SUMMARY

Whey and whey permeates are byproducts of cheese making. After removal of casein and fat from whole milk, a product arises which contains practically all the lactose of the milk. The low protein content of the product precludes whey from being accepted as a high-grade food or feed. Because of its high water content, the transport of this liquid is expensive. So the products have to be dried and to be sold at a price that would offset the drying costs. Even worse is the situation with whey permeate, which is less useful for feed and is difficult to be dried. The lactose of whey has been proposed some 40 years ago as a carbohydrate substrate for biotechnological transformations, especially for food- and feed yeast production. General aspects of technical processes and results of pilot plant studies are discussed in this review.

## 1. Introduction

There is a tendency in the dairy industry to concentrate its plants in a few locations, while the capacity of plants is growing caused by this concentration and by increasing consumption. This is true especially for cheese making. In general cheese consumption in the developed countries increased for about 50 % and more in 15 years (Table 1). The mean consumption in these countries is 10,9 kg per capita and year.

The whey problem is mainly a problem of developed countries since cheese consumption in developing countries is low, it is for Africa, South America and Asia 0,44 kg per capita and per annum (Table 2). So the construction of whey processing plants is no matter of export to countries of the third world.

Cheese is a high-price product in food industry. The quantity of vat milk, which is necessary to produce one kg of cheese depends on the sort of cheese to be processed (Table 3). For Emmentaler cheese 12,5 kg vat milk is needed for 1 kg cheese and approximately 11,3 kg sweet whey results after treatment with rennet. For quark production with starter cultures the yield of whey is about 4 kg acid whey. In line with the need of vat milk is the increased concentration of casein in cheese. Assuming !n average yield of 8 kg whey per kg cheese, the world production of whey in 1981 was 92,8 millions tons.

## 2. Components of whey for yeast fermentation

Yeast fermentation of whey may yield single cell protein and/or ethanol. Since casein and fat go nearly quantitatively into the cheese, the main carbon sources for fermentation are lactose and lactic acid. For ethanol production, only lactose is used by the yeast in the glycolytic pathway, while for oxidative single cell production, both kinds of lactic acid, L- and D-isomers are metabolised also. So only the sweet whey or ultrafiltration permeate from cheese manufactured with rennet is suited for ethanol production and care has to be taken, that sweet whey does not get sour by the activity of lactic acid bacteria during storage before ethanol fermentation. Both, the acid whey and sweet whey may be used for single cell production, since yeasts are known to use the lactic acid quantitatively. The quantity of lactic acid may exceed 0,8 % (Table 4) considerably. For our own technical experiments with single cell production in Hameln and in Flensburg (West Germany), a whey was delivered to us from the Harzer cheese-manufacture with 1,93 % lactic acid and a pH of 3,5. Such a whey is not suited for ethanol production. Traces of fat may pass the separator and may get into the whey. This fat may act as an anti-foam agent and has in this property a favourable influence on the fermentation process. Non-casein-protein, which can not be coagulated by rennet or by lactic acid, is left in the whey. Globulins and Albumins are valuable proteins for food. They may be separated from the whey by heat treatment or by ultrafiltration before further processing the whey and may be added to the cheese. From table 4 we learn, that half of the milk dry matter is still found in the whey.

Laying out a process for yeast fermentation from whey it has to be taken into account that whey processing is influenced in future by the above mentioned membrane-separation technics (Table 5). While the original whey may be used for direct feeding, there are problems to use permeate after separation of soluble whey proteins or of coagulated whey protein after heat treatment. So permeate is another substrate in question for yeast fermentation. After lactose production from concentrated whey or permeate, molasses are formed which are similar in some way to molasses of sugar processing from sugar beets.

Lactose-molasses from concentrated whey contain 50-55 % dry matter, 22-28 % lactose, minerals and protein 22-26 %. Since all the minerals and all of the protein from whey comes into the molasses, the relation minerals: protein is nearly the same as in the whey, approximately 1 : 1. So for yeast fermentation a high mineral content of molasses has to be taken into consideration.

Modern processes of yeast fermentation with lactose as carbon source are based on deproteinized whey as substrate. From 1 t of whey 5 kg highly valuable protein of food grade can be obtained by ultrafiltration. If large quantities of whey are available, it is advisable to separate this protein before fermentation: Single cell protein production additionally yield at best 25 kg biomass dry matter/ 1 t of whey with 50 % protein that is 12,5 kg protein/t of whey, which is in mixture with RNA, carbohydrates and fats. Enrichment procedures are necessary to reach the quality of pure whey protein.

## 3. Biomass production

In table 6 Food- and Feed-grade yeast plants are listed up. In World War II Mietke and Dubrow (1944) reported that 8 yeast factories were operating in Germany based on the substrate whey. The maximal capacity for each was up to 800 tons/year. The production in that time was limited by transport mainly. Two plants based on whey had been installed in the United States, another process of SAV had been operated in France, Russia and Czechoslowakia on a small scale. A plant producing Food-grade yeast from deproteinized whey is operation in Vendome, France, producing approximately 10 000 tons SCP/year.

Another plant, which is in operation in the Irish Republic, using permeate from whey as a substrate for ethanol production with yeast can produce 4 000 tons Ethanol/year (Table 7). 600 t of whey per day must be available in the location of production. For cheddar cheese production - cheddar is the main product in Carbery-approximately 660 t of milk have to be supplied to the plant to yield 600 t of whey. This milk supply is exceptionally high. In Western Germany only 5 dairies out of 720 have a milk supply of 33 t/day and less (1979, BML-study). Smaller dairies will have to consider other possibilities of whey utilisation without applying membrane separation technics or separation of protein by heat coagulation. One way is SCP production. In table 8, different processes of SCP production from whey are represented, which are more or less classical in the layout. One way is a yeast fermentation of whole whey, succeeded by concentrating the biomass in the evaporator and spray or roller drying of the total product. In this case, no waste water is produced, except vapours of the evaporator with some organic matter. A disadvantage is: Total drying of biomass with a high content of water (96% $H_2O$) is expensive. Another way is to separate the yeast after fermentation putting up with the production of waste water. In this case, a special heat treatment of whey protein must be included in the process. Heat coagulation of protein before yeast fermentation and separation of protein together with the yeast leads to technical difficulties, for the coagulated protein may settle in the tubings. Heat coagulation after fermentation has some advantage since the yeast is inactived simultaneously. A disadvantage of the variant is, that protein and other substances of the yeast diffuse from the cells into the water and the $BOD_5$ of the waste water increases. In the fromagerie-Bel-process deproteinized whey is fed to the fermenter and yeast is separated after fermentation without difficulties.

In table 9 the standard description of yeasts used in technical and/ or in pilot plant processes with whey are represented. For ethanol fermentation Kluyveromyces fragilis and lactis, Candida pseudotropicalis and kefyr are able to ferment lactose, these species had been the basis for a screening programme in Ireland. For SCP-production from whey, those yeasts may be preferred which can assimilate lactose, but cannot ferment lactose, since production of ethanol beside SCP-production may lead to a loss of yield. Yeast of this kind is C.intermedia. Beside lactose assimilation, the assimilation of lactic acid in acid whey, of citric acid and the need of vitamins for growth has to be taken into consideration. In our pilot plant studies we used a mixed culture of C.krusei and Lactobacillus bulgaricus for fermentation. The Candida uses the Lactic acid produced by the lactic acid bacteria. In order to spare cooling water those yeasts may be preferred, which have a high maximum temperature of fermentation or of growth (above 40° C).

In our experiments, Lembke et al. (1975) and in those of Bayer and Meyrath (1979) only nitrogen had to be added to the fermentation, phosphorus and vitamins were supplied sufficiently with the whey (Table 10). In order to omit heat treatment of whey protein we used yeast-own proteinases and peptidases for enzymatic degradation of whey protein, which are liberated from yeast cells by autolyses caused by osmotically active substances. For preparation of the autolysate, 10 kg of yeast harvest and 10 kg of ammoniumsulphate were mixed and held 24 hours at 40° C. This preparation was added to 4 t of whey. No further addition was made.

The lay-out of the process has an important influence on composition of biomass. Using whey as the only carbon source and drying the total product, the percentage of crude protein in the dry matter is low and the percentage of minerals is high (Table 11). All the minerals of the whey are included in the product and additional salts are introduced by addition of sulphate and by neutralisation of sulphuric acid which is left in the broth after assimilation of ammonia by the yeasts. In practive even 20 % of minerals and more are analysed after drying the total biomass. In the food law of the Federal Republic of Germany only 8 % minerals in yeast dry matter are permitted.

A product with low mineral content and high protein content in case of drying the total product may be obtained, if carbon sources with low mineral content are added to the whey. Forman et al. (1975) added ethanol to the process. Moebus et al. (1979) added enzymatically hydrolysed starch to the fermenter. In the latter process the lactose of the whey was converted into a mixture of lactic acid/ammoniumlactate by lactic acid fermentation. This mixture and the glucose from starch could be fermented with a normal baker's yeast. After drying of the fermented material, the crude protein was 49 % and the ash content was 5,7 %. The yields of whey fermentation are in the region of 50 % for SCP. The theoretical value for ethanol production is 53,8 % ethanol/100 g lactose. In practice (Carbery process) 45 % ethanol are obtained (Table 12).

The reduction of the Biological Oxigen Demand ($BOD_5$) is in the same order for waste water from ethanol production and from SCP-production (with separating the yeast from the fermentation broth with the centrifuge). The lowest values for the $BOD_5$ are in the order of 5-10 % of the original whey/permeate (Table 13).

It is seen from table 14, that these values may be compared with effluents of the dairy industry in general, which have to be treated in a purification plant.

4. Objects of SCP-production and economics in the future

SCP-production from whey cannot compete with soyameal since this is cheaper and plentiful. From the marketed products from the actually working plants the following conclusion can be drawn: If single cell proteins are used for feeding, a product can be sold, which is high in protein content, so it may be used for upgrading low protein feed rations, especially in milk replacer for calves (Table 15). Another use of these high protein SCP products could be found in the food industry, as special additives for products, warranting special functional properties.

For dairies with a smaller milk supply it is more favourable to use supplemented whey, for instance with hydrolysed starch. A comparison of production costs, evaluated by Vogelbusch (Vienna) for a process with Candida intermedia and acid whey (pilot plant in Klagenfurt, Austria) and a process with acid whey + wheat flour (pilot plant in Flensburg, FRG) shows that even an addition of an expensive carbon source results only in a small increase of production costs (Table 16).

Future opportunities for SCP-production on the basis of different substrates were summarized by Hepner (Table 17). n-alkanes and gas-oil seem to be uneconomic, except in COMECON-countries. Methanol would be viable, if derived from non-petroleum feed-stock. Ethanol as a substrate will be economic, if SCP is sold as a food ingredient. More plants may be built on the basis of sulphite liquor. Starch as a renewable source may be used for SCP production. But if fermentation ethanol is subsidised by European governments in the 1980's, it may become a more attractive product than SCP. Under these circumstances the demand for starch (maize) could exceed its availability and could, on the other hand, force up soya prices in the 1980's, since US lands would be increasingly used for the cultivation of maize.

REFERENCES

1. BAYER, K. and MEYRATH, J. (1979) Biomass from whey in Economic Microbiology 4, 207, Academic Press, London
2. FAO production yearbook (1982) 35, 61-71, 234-235, Rome
3. FORMAN, L. et al. (1975) Kvasný průmysl, 21 (12) 283

4. HEPNER, L. (1980) SCP - The past decade and future opportunities in Microbielle Proteingewinnung und Biotechnologie 2. Symposium, 1 Verlag Chemie, Weinheim, 1982
5. LEMBKE, A. et al. (1975) Versuche zur Herstellung von Einzellerprotein aus Molke in einer halbtechnischen Versuchsanlage, 571, Sdrh.Ber.Ldw.192
6. LODDER, J. (1970). The yeasts. North Holland Publishing Comp. Amsterdam
7. MIETKE, M. and DUBROW, H. (1944). Der heutige Stand der Molkenverwertung. Deutsche Molkerei- und Fettwirtschaft 37, 290
8. MOEBUS, O. (1976). Einzellereiweiß aus Molke und Permeaten. Die Molkereizeitung "Welt der Milch", 30, 31, 897-898, 900-902
9. MOEBUS, O. and TEUBER, M. (1979). Herstellung von Single Cell Protein mit Saccharomyces cerevisiae in einer Tauchstrahlbegasungsanlage. Kieler Milchwirtschaftl. Forschungsberichte 31, 297
10. NAIDITCH, V. and DIKANSKY, S. (1960) cit. L. L. Muller, Progress Biochemistry 4, 21, 1969
11. NIEMEYER, H. Handbuch für Molkereifachleute (1959). 388, Verlag Th. Mann GmbH, Hildesheim
12. POWELL, M. E. and ROBE, K. (1964). Food Process 25, 80, cit. Mateles et al. Single cell protein 237 The M.I.T. Press, Cambridge, 1968
13. REESEN, L. (1978). Source: Dairy Industries International, Facts about whey alcohol 43; 1; 9, 16
14. SCHWARTZ, R. D. and LEATHEN, W. W. (1976) cit. Alkinson, B. et al. Biochemical Engineering and Biotechnology Handbook 1058, The Nature Press, New York, 1983
15. THIESSEN, U. (1982). Alkoholgewinnung aus Molke. Fette, Seifen, Anstrichmittel 84, 4, 164
16. VOGELBUSCH (1978). Wirtschaftlichkeitsstudie. Analyse und Entsorgungssituation der Nahrungs- und Genußmittelindustrie (Milchverarbeitung) in der Bundesrepublik, Vienna
17. VRIGNAUD, Y. (1976) Lactose-Hefen. Die Methoden der Produktion und ihre Verwertung in der Ernährung. Milchwirtschaftliche Berichte, Wolfpassing und Rotholz 50, 1, 1977
18. WIRTHS, W. (1974). Kleine Nährwerttabelle der Deutschen Gesellschaft für Ernährung e.V. 8. Im Umschau-Verlag, Frankfurt a. M.

TABLE 1

ANNUAL PRODUCTION OF CHEESE 1981 (FAO 1982)
Developed countries

| | production | | increase (1961-1971) =100 |
|---|---|---|---|
| world | 11,6 | $10^6$ Metric tons | 51,4% |
| NC America | 2,59 | $10^6$ Metric tons | 64,3% |
| Europe without USSR | 5,78 | " " " | 53,6% |
| USSR | 1,51 | " " " | 50,3% |
| Australia | 0,137 | " " " | 79,2% |
| Japan | 0,071 | " " " | 76,2% |
| South Africa | 0,031 | " " " | 56,3% |
| total | 10,13 | $10^6$ Metric tons | |

Cheese consumption: 10,9 kg per capitum and per annum

(924 Mill. people)

TABLE 2

ANNUAL PRODUCTION OF CHEESE 1981 (FAO 1982)
Developing countries

| | | | |
|---|---|---|---|
| Africa w/o South Africa | 0,339 | $10^6$ Metric tons | 23,5% |
| South America | 0,457 | " " " | 21,1% |
| Asia w/o Japan | 0,617 | " " " | 40,7% |
| total | 1,41 | $10^6$ Metric tons | |

Cheese consumption: 0,44 kg per capitum and per annum
(3208 Mill. people)

TABLE 3

YIELD OF WHEY

| SORT OF CHEESE | KG VAT MILK | YIELD/KG CHEESE | % PROTEIN | %PROTEIN CHEESE / %CASEIN MILK (=2,8) |
|---|---|---|---|---|
| EMMENTALER (HARD CHEESE) | 12,5 | 11,3 KG SWEET WHEY | 27 | 9,6 |
| EDAMER (SEMIHARD CHEESE) | 11,0 | 9,7 KG SWEET WHEY | 24 | 8,6 |
| CAMENBERT (SOFT CHEESE) | 8,7 | 7,5 KG SOUR WHEY + RENNET | 19 | 6,8 |
| COTTAGE CHEESE, QUARG (FRESH CHEESE) | 5,3 | 4,0 KG SOUR WHEY | 17 | 6,1 |

TABLE 4

| COMPONENTS OF MILK AND WHEY | COW'S WHOLE MILK % (W/W) | SWEET WHEY % (W/W) | SOUR WHEY % (W/W) |
|---|---|---|---|
| LACTOSE | 4,9 | 4,9 | LACTOSE 4,0<br>LACTIC ACID UP TO 0,8 |
| FAT | 3,9 | 0 | 0 |
| CASEIN | 2,8 | 0 | 0 |
| LACTO-GLOBULIN | 0,30 | CRUDE PROTEIN 0,9 | CRUDE PROTEIN 0,9 |
| LACTO-ALBUMIN | 0,15 | | |
| NON-PROTEIN N×6,38 | 0,12 | | |
| MINERALS | 0,7 | 0,6 | 0,75 |
| WATER | 87,1 | 93,6 | 93,5 |
| DRY MATTER | 12,87% | 6,4% | 6,5% |

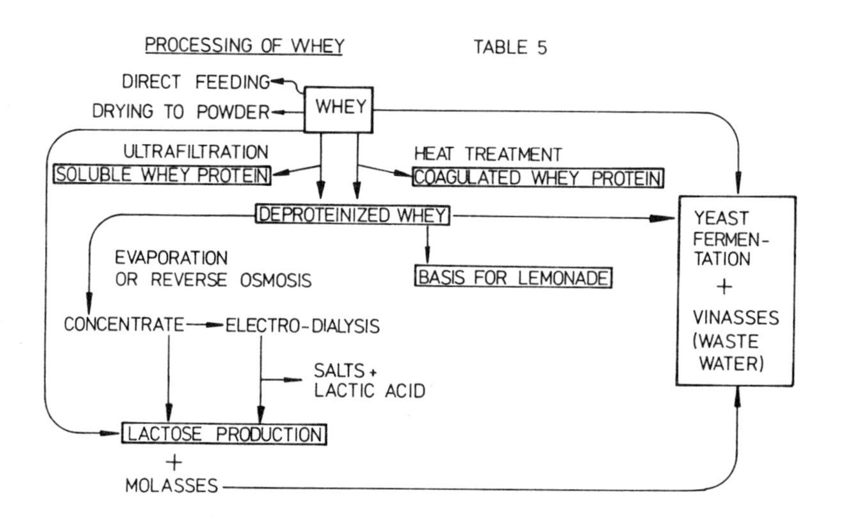

TABLE 6 FEED-GRADE YEAST PLANTS

| PLANT LOCATION | PRODUCTION (TONS/YEAR) | REFERENCE |
|---|---|---|
| GERMANY | 8 YEAST FACTORIES, CAPACITY EACH UP TO 800 | MIETKE & DUBROW 1944 |
| FRANCE (VENDOME) | 10000 | VRIGNAUD 1976 |
| USA (MILBREW) | 5000 | SCHWARTZ & LEATHEN 1976 |
| USA (KNUDSEN) | 800 | POWELL & ROBE 1964 |
| CZECHOSLOVAKIA (SAV) | 300 | NADITCH & DIKANSKY 1960 |
| FRANCE " | — | " |
| RUSSIA " | — | " |

TABLE 7

PRODUCTION OF FOOD-GRADE AND TECHNICAL ALCOHOL FROM WHEY PERMEATE

Company : Carbery Creameries
Plant Location: Irish Republic
plant size : 4.000 tons Ethanol/year
cited : U. Thiessen (1982)

TABLE 8 SCP-PRODUCTION FROM WHEY

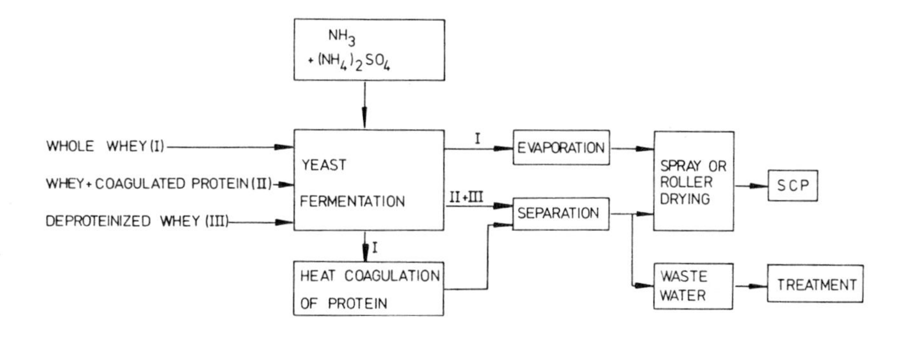

TABLE 9

STANDARD DESCRIPTION OF YEASTS (LODDER 1970) USED FOR FERMENTATION

| YEAST | FERMENTATION OF LACTOSE TO ETHANOL | ASSIMILATION OF LACTOSE | ASSIMILATION OF LACTIC ACID | ASSIMILATION CITRIC ACID | VITAMIN FREE MEDIUM | GROWTH TEMPERATURE |
|---|---|---|---|---|---|---|
| K.FRAGILIS | + | + | + | + OR − | NO GROWTH | AT 37°C POSITIV |
| C. PSEUDO-TROPICALIS | + | + | + | +(WEAK)OR− | NO GROWTH | MAX. 44−47°C |
| K. LACTIS | + | + | + OR − | − | NO GROWTH | NO GROWTH AT 37°C OR VERY SLOW GROWTH |
| C. INTERMEDIA | − | + | − | − | WEAK GROWTH | MAX. 29−30°C |
| C.KEFYR | + | + | + OR WEAK | − | NO GROWTH | MAX. 37−42°C |
| C.UTILIS | − | (AFTER ADAPTION + ?) | + | + | WEAK TO GOOD GROWTH | MAX.39−43°C |

TABLE 10

NUTRIENT REQUIREMENTS

1.) Nitrogen, in form of ammonium compounds, has to be added
2.) Phosphorus is supplemented in sufficient quantity with the whey
3.) The vitamin requirements vary from strain to strain
   a) Continuous cultures with Cd. intermedia performed with high dilution rates and high yields without addition of vitamins (Bayer, Meyrath 1979)
   b) No vitamins are added in the process with Cd. krusei and Lb. bulgaricus (Lembke et al. 1975)

TABLE 11

COMPOSITION OF BIOMASS

| SUBSTRATE | RECOVERY OF BIOMASS | %CRUDE PROTEIN DRY MATTER BASIS | %MINERALS DRY MATTER BASIS | PROCESS |
|---|---|---|---|---|
| WHEY AS ONLY CARBON SOURCE | TOTAL PRODUCT DRIED | 1) 25−36 | 16 − 20 | SAV - PROCESS |
| | | 2) 41,9 | 16,7 | BAYER,K. AND MEYRATH,J. (1976) |
| | YEAST SEPARATED W/O WHEY PROTEIN | 47 | 7 | BEL-PROCESS VRIGNAUD,Y. (1976) |
| WHEY + ADDITIONAL CARBON SOURCES 1t WHEY + 0,2t WHEAT FLOUR | TOTAL PRODUCT DRIED | 49 | 5,7 | MOEBUS, O., TEUBER, M. (1979) |
| | YEAST SEPARATED W/O WHEY PROTEIN | 51 | 5,8 | MOEBUS, O., TEUBER, M. (1979) |

TABLE 12

YIELDS REPORTED IN WHEY FERMENTATION (BAYER ET AL.1979)

| AUTHOR/PROCESS | ORGANISM | YIELD PER 100 g LACTOSE |
|---|---|---|
| BEL PROCESS VRIGNAUD | K.FRAGILIS | 50 g SCP |
| LEMBKE ET AL. | C. KRUSEI AND LB. BULGARICUS | 52 g SCP |
| SAV | K. FRAGILIS | 36−42 g SCP |
| BAYER ET AL. | C. INTERMEDIA | 54 g SCP |
| CARBERY | LACTOSE YEASTS | 45 g ETHANOL |

TABLE 13

RESIDUAL BIOLOGICAL OXIGEN DEMAND $BOD_5$
AFTER FERMENTATION OF WHEY

SUBSTRATE WHEY: 40 000–50 000 mg $O_2$/l

| PROCESS/AUTHOR | MICROORGANISMS | $BOD_5$ (mg $O_2$/l) |
|---|---|---|
| CONTINUOUS ETHANOL-PRODUCTION L. REESEN (1978) | K. FRAGILIS | 4500 |
| CONTINUOUS SCP-PRODUCTION FROMAGERIE BEL V. VRIGNAUD (1976) | K. FRAGILIS AND K. LACTIS | 1300–1500 |
| SCP-PRODUCTION BATCH (INFERATOR) A. LEMBKE ET AL. (1975) | MIXED CULTURE: C. KRUSEI +LB. BULGARICUS +ACTIVE YEAST AUTOLYSATE | 2200 |

TABLE 14

APPROXIMATE VALUES FOR BOD IN
WASTE WATERS O. MOEBUS (1976)

| KIND OF WASTE WATER | BOD: 5 DAYS |
|---|---|
| DAIRY INDUSTRY FOR EACH 1L MILK 0,8–1,5 L WASTE WATER | ~1400 MG $O_2$/L |
| HOUSEHOLD | 400 |
| TREATED WASTE WATER | 20 |

Table 15

OBJECTS OF SCP-PRODUCTION

For feeding

1. Product with a high protein content for upgrading low protein feed rations (milk replacer for calves)
2. Addition to the rations of dogs, cats and birds in small quantities (3-7%)

For food industry

1. Product with low RNA-content
2. Product with special functional properties:
   Water binding, fat binding
   dispersibility, emulsification
   gelation, thickener
   texturisation, fiber formation
   flavor, flavor enhancing
   dough elasticity, leavening (baker's yeast)
   heat coagulation
   whipping

TABLE 16

COMPARISON OF PRODUCTION COSTS FOR YEAST PRODUCED FROM WHEY BASED ON A DAILY SUPPLY OF 260.000 l OF WHEY

| | whey | supplemented whey |
|---|---|---|
| production | 3.300 t/yr | 12.850 t/yr |
| costs | 1,569 DM/t | 1.750 DM/t |
| analysis: | | |
| whey | 28,8 % | 6,6% |
| flour | - | 45,8 |
| chemicals | 5,7 | 6,7 |
| electrical energy | 6,0 | 5,0 |
| steam | 17,4 | 9,4 |
| cooling water | 2,1 | 3,9 |
| labour | 24,4 | 11,4 |
| div. | 7,1 | 7,3 |
| repairs & pay off | 7,4 | 3,9 |

source: Vogelbusch GmbH, Vienna (1978)

TABLE 17

FUTURE OPPORTUNITIES FOR SCP (HEPNER, 1980)

| SUBSTRATE | SOURCE | PROSPECTS FOR SCP IN THE 1980'S |
|---|---|---|
| N-ALKANES AND GAS-OIL | FOSSIL | UNECONOMIC, EXCEPT IN COMECON COUNTRIES |
| METHANOL | FOSSIL OR RENEWABLE | COULD BE VIABLE, IF DERIVED FROM NON-PETROLEUM FEEDSTOCK |
| ETHANOL | FOSSIL OR RENEWABLE | ECONOMIC, IF SCP IS USED AS A FOOD INGREDIENT |
| SULPHITE LIQUOR | RENEWABLE | FEASIBLE, MORE PLANTS MAY BE BUILT |
| STARCH | RENEWABLE | IF FERMENTATION ETHANOL IS SUBSIDISED, IT MAY BE MORE ATTRACTIVE FOR ETHANOL-PRODUCTION |
| WHEY | RENEWABLE | FEASIBLE, PROVIDED SUFFICIENT QUANTITIES ARE AVAILABLE IN ONE LOCATION |

# UTILISATION OF WHEY AND ULTRAFILTRATION PERMEATES

S G COTON
Milk Marketing Board

Summary

A number of current uses for whey and potential commercially interesting new uses for whey are described and cost models are presented on the common basis of return per litre of whey. This enables comparison between the various uses and comparison with the production of single cell protein from whey.

My subject does not fit into the main stream of this Workshop's consideration of production and feeding of SCP. The purpose of the paper is to describe and expose the alternatives to production of SCP from whey and permeate derived from whey ultrafiltration so that production of SCP may be placed in context with these other options. To do this properly requires consideration of the costs and benefits deriving from alternative uses of whey and I touch upon economic aspects whenever possible though it is very difficult to give good cost models, in part because of technical difficulties, in part because where figures are accurately known they are often confidential and thirdly because economics vary from time to time and from country to country and indeed from company to company within a country depending upon particular circumstances. (see figure 1)

In order to commence my subject I should try to define the scale of the topic. Figure 1 shows the fate of the various milk solids in the course of making cheese. The point of consequence from this figure is that exactly half of the solids in milk remain in the by-product of cheese manufacture - whey. Finding profitable utilisation for these whey solids must therefore be a constant preoccupation of the dairy industry.

Table I gives the composition of whey from cheeses such as Cheddar.

Table I - Average Composition of Sweet Whey

| Component | Percentage |
|---|---|
| Lactose | 4.8 |
| Protein | 0.7 |
| Non Protein N | 0.2 |
| Ash | 0.6 |
| Fat | 0.05 |
| Lactic Acid | 0.15 |
| Total Solids | 6.50 |

One thing stands out with great clarity - lactose is by far the major constituent. Indeed it dominates to the point where this paper could, with justification have been entitled "Utilisation of Lactose", a subject of fundamental importance to the dairy industry.

Very large quantities of lactose are available in by-products of the dairy industry, skimmed milk and whey. Confining our attention to whey, available solids are shown in Table II.

Table II - Total Whey Solids Availability

| | 1981<br>k Tonnes/Year |
|---|---|
| EEC | 2,525 |
| Other W Europe | 380 |
| Eastern Europe | 1,785 |
| N America | 1,628 |
| S America | 281 |
| Australasia | 156 |
| Other Countries | 932 |
| | 7,687 |

(Source: MMB "Facts & Figures" 1982)

Of the 7.7M tonnes of whey solids available about 4.9M tonnes is lactose. This compares with a world production of about 90M tonnes of sucrose. Quantities of whey solids, or lactose, are therefore quite small but still significant in world terms compared with the major sugar sucrose. Unfortunately, however, cheese manufacture by its very nature is a widely dispersed industry and at any one site there are relatively small quantities of whey solids available. These quantities are often insufficient to justify the capital necessary for sophisticated processing; for example, a very large Cheddar cheese factory will produce no more than 10,000 tonnes/year of whey solids.

I now consider the options for whey processing:

## CURRENT UTILISATION

Statistics on utilisation of whey are not readily available in many parts of the world. In the EEC where the value of whey has for long been recognised, some statistics are available showing the following:

Table III - Whey Utilisation in EEC

| | | % |
|---|---|---|
| 1. | Liquid to stock feeding | 45 |
| 2. | Whey Powder | 30 |
| 3. | Lactose and delactosed whey powder | 15 |
| 4. | Others | 5 |

To consider each of these in turn:

### Stock Feeding

Pig rearing has been traditionally associated with the dairy industry in a number of countries, particularly in Denmark and in our host country, Switzerland, over 90% of whey production is used for pig feeding. Liquid whey, however, containing over 90% water is an expensive material to transport and, after allowing for the cost of transport, profit to the dairy from selling whey for pig feeding can be very small and, with the increase in size of cheese creameries, it becomes increasingly difficult to rely on pig feeding as an outlet for whey.

## Whey Powder

In the USA whey powder finds considerable use in human food. This pattern is not reflected in Europe where by far the major use is for incorporation into animal feeds. Currently, the financial return on manufacture of whey powder is quite good within EEC but the market fluctuates considerably and in some recent years financial return has been zero or even negative. Table IV gives some indication of costs and returns for whey powder.

Please note that in this and subsequent costings capital and revenue costs are incorporated into the unit process costs and profit/loss is related to the unit of one litre of whey so that comparisons can be made on a common basis. It is assumed that the whey has zero value before processing; in some circumstances it may however have a negative value, for example when it is discharged to an effluent plant.

Table IV - Whey Powder Cost Model

| | £/tonne | |
|---|---|---|
| Evaporation | 92 | |
| Spray Drying | 64 | |
| Packaging | 20 | |
| Storage/Marketing/Transport | 35 | |
| | | 211 |
| Suggested Selling Price | | 275 |

Profit = £64/tonne of whey (0.43 pence/litre whey)

## Lactose and Delactosed Whey Powder

Estimated production of lactose is shown in Table V:

Table V - Actual World Lactose Production

| Country | k Tonnes |
|---|---|
| US (1974) | 58 |
| New Zealand (1974) | 10 |
| Netherlands (1976) | 62 |
| Germany (1976) | 35 |
| Great Britain (1979) | 12 |
| France (1976) | 10 |
| Ireland | 1 |
| | 188 |

Note: E European production not known
Source: Fox, P F (1980) J SDT 33, 118 with additions

Total production of 188k tonnes lactose compared with the total available from whey of 4.9M tonnes is very small, in world terms amounting to only 3.8% though as my previous Table has indicated lactose production is of greater consequence within EEC.

The largest single use of lactose is in the modification of cows milk to simulate human milk in infant food formulations and the second largest use in pharmaceuticals where lactose is extremely useful in tableting. Within the USA there are some useful statistics on the end of lactose which are reproduced in Table VI.

Table VI - End Use of Lactose, USA (1978)

| Use | Percentage |
|---|---|
| Infant Foods | 50.2 |
| Pharmaceutical | 13.4 |
| Dietetic Foods | 12.2 |
| Dairy Products | 10.9 |
| Other | 13.3 |

Source: Allum D (1980) J SDT 33, 59

The cost of producing lactose is shown in Table VII:

Table VII - Refined Lactose Production Cost Model

| | £/tonne | |
|---|---|---|
| Evaporation | 260 | |
| Extraction and Drying | 20 | |
| Chemical Treatment/Recrystallisation | 60 | |
| Packaging | 20 | |
| Marketing | 30 | |
| | | 390 |
| Suggested Selling Price | | 380 |

Loss £10/tonne (0.02 pence/litre whey)

You will see that in present economic conditions this operation is likely to result in a loss.

The production of refined lactose inevitably results in the production of a by-product, the mother liquor from which the lactose is crystallised and this is commonly dried and sold as delactosed whey. A cost model for the production of delactosed whey is given in Table VIII.

Table VIII - Delactosed Whey Cost Model

| | £/tonne | |
|---|---|---|
| Drying | 161 | |
| Packaging | 20 | |
| Marketing | 20 | |
| | | 201 |
| Suggested Selling Price | | 270 |

Profit £69/tonne (0.11 pence/litre whey)

Under present economic conditions you will see that I estimate this to result in a modest profit.

Putting the two products lactose and delactosed whey together results in the following:

Table IX - Lactose Plus Delactosed Whey Cost Model

| | pence/litre whey |
|---|---|
| Lactose Extraction | (0.02) |
| Delactosed Whey Powder | 0.11 |
| Profit | 0.09 |

The net consequence of lactose and delactosed whey production is shown in this Table to be a very modest profit.

Whilst a very large number of potential uses of lactose have been identified, it seems unlikely that world demand for lactose will increase to any significant extent. The dairy industry has therefore focused its attention upon other uses for whey and the lactose within whey and it is these more modern developments to which I would like to give attention.

## RECENT AND FUTURE USE OF WHEY

### Protein Extraction - Ultrafiltration

Ultrafiltration (UF) has two major uses in the dairy industry both related to the manufacture of cheese; firstly to extract protein from whey and secondly to concentrate the total protein and fat in milk prior to cheesemaking. This second process results in the whey proteins being incorporated in the cheese and has thus far been developed to a commercial scale for soft cheese. There are increasing prospects for its use in the manufacture of hard cheese and this would dramatically increase the total of membrane processing which has recently shown remarkable growth as exhibited by Figure 2.

The total world capability for extracting protein from whey is probably now greater than the current market for whey protein and I predict a marked slowing of growth in this area but a continuing increase in use of UF prior to cheesemaking.

Whey proteins have a number of established and potential uses in baby foods, meat processing and as partial replacement for eggs in baking. The UF process is capable of producing a range of whey protein products up to about 75% protein in total solids.

A model costing for protein extraction from whey is given in Table X.

Table X - UF Protein (60%) Cost Model

| | £/tonne | |
|---|---|---|
| Whey Pretreatment | 245 | |
| UF Extraction | 950 | |
| Spray Drying | 318 | |
| Packaging | 20 | |
| Marketing | 50 | |
| | | 1583 |
| Suggested Selling Price | | 1700 |

Profit £117/tonne (0.11 pence/litre whey)

Protein Extraction - Ion Exchange

A disadvantage of UF applied to whey is that is is limited to producing a product with a maximum protein content of 75% of total solids and fat is concentrated to the same extent as protein. This limits applications, particularly those which require the best possible foam production. Whey proteins can also be extracted by ion exchange techniques; the English company, Bio Isolates, and the French company, Rhone Poulenc, have each developed an ion exchange process to a potential commercial scale. The ion exchange processes give products with no fat and greater than 90% protein in dry matter. A cost model for their manufacture and sale is given in Table XI.

Table XI - Ion Exchange Protein Cost Model

| | £/tonne | |
|---|---|---|
| Whey Pretreatment | 450 | |
| Ion Exchange Extraction | 700 | |
| UF Concentration | 900 | |
| Spray Drying | 318 | |
| Packaging | 20 | |
| Marketing | 50 | |
| | | 2438 |
| Suggested Selling Price | | 3000 |

Profit £562/tonne (0.29 pence/litre whey)

Permeate

Ultrafiltration and ion exchange of whey leaves a lactose rich 'permeate' and its utilisation is essential to ensure economic viability for protein extraction. Much attention has been devoted to this utilisation and one of the most promising possibilities is enzymic hydrolysis of the lactose to its constituent monosaccharides, glucose and galactose. We have collaborated with the Corning Glass Company to develop their process of enzyme immobilisation to the hydrolysis of lactose in whey permeate and in whey. The system utilises a B-galactosidase immobilised on to silica beads by use of a silane-gluteraldehyde linkage and is capable of hydrolysing lactose in both whey and whey permeate and thereby enables these by-products to be used in a number of food applications for which they would otherwise be unsuitable.

The process has now been developed to the point that the Milk Marketing Board and the Corning Company have jointly set up a company called Specialist Dairy Ingredients which currently has a small commercial/very large pilot plant operating the hydrolysis process in association at one of the MMB's cheese plants in Aston in Shropshire. Planning is well advanced for the building of a very large commercial plant.

The process was first developed for use on whey permeate and Figure 3 gives an outline of the basic process as applied to permeate. The permeate, which should have a true protein content of not greater than 0.05% is pasteurised and then demineralised using a twin column ion exchange system though, depending upon conditions, electrodialysis or a combination of electrodialysis and ion exchange may be preferred. Permeate is then adjusted to pH 4.5 by the addition of 5% HCl and cooled accurately to the operating temperature of the enzyme column through which it is pumped in downward flow at a controlled flow rate. During passage through the column and its enzyme bed hydrolysis of the lactose occurs to the extent of 90-95% conversion to glucose and galactose. The plant is operated at constant throughput which requires a very gradual raising of the temperature of operation from 32°C to about 50°C to compensate for the slow loss of enzyme activity. After hydrolysis, the permeate is evaporated to a syrup of 60-68% total solids when it has a sweetness similar to medium DE corn syrup compared with which it is significantly less viscose at ambient temperature and corresponding solids. Filtration is usually necessary to give a bright, sparkling end product and where the application demands, lime and activated carbon treatment, can be applied prior to filtration to give a water-white product which corresponds in visual quality to top grade corn syrup.

(see figure 4)

The process has now been developed to operate entirely satisfactory on whole whey which greatly extends the possibilities of its application both in the sense that the process is not dependant upon the existence of whey ultrafiltration plants and the range of applications of hydrolysed whey are greater than those of hydrolysed permeate. The process is generally similar to that which I described for permeate, except that demineralisation takes place after hydrolysis and a proprietory process step is required between the pH adjustment/pasteurisation and the hydrolysis column in order to prevent any deposition of denatured protein in the enzyme bed.

A major achievement in this particular process is that cleaning and sanitisation of the enzyme bed is simple involving circulation of a proprietory cleaning agent formulated to remove protein deposits from the enzyme bed followed by circulation of a dilute solution of a bactericide and then by dilute acetic acid pH 4.

Demineralisation is a significant proportion of the total cost of the process and the degree of ash removal is consequently tailored to the particular end use of the hydrolysed permeate or whey.

The two most interesting applications for hydrolysed whey are in ice cream and sugar confectionery though these are by no means the only outlets. Hydrolysed whey can be used in any product where the combination of milk protein together with a comparatively sweet sugar is required.

We have shown that satisfactory ice cream can be produced containing as much as 20% of their solids as hydrolysed whey syrup. The hydrolysis of lactose in whey not only makes the lactose itself more acceptable in this application but by doing so permits the use of whey in ice cream at much higher levels than otherwise would be the case. This brings advantages of both texture improvement and reduction of ingredients costs. This application is an instance where there is no necessity to fully demineralise; demineralisation of somewhere between 50 and 70% is probably adequate. In general, ice cream becomes softer as the level of hydrolysed whey is increased, which can be an advantage or disadvantage depending on the texture which is required.

A particularly promising application of hydrolysed whey is in the manufacture of soft toffees for which a blend of whey concentrate and corn syrup is usually deliberately prepared. If the lactose in the whey concentrate is hydrolysed the need to buy corn syrup to produce the required blend is obviated and the amount of whey solids per tonne of blend replacement can be approximately doubled. Once again, this is an application which the degree of demineralisation can be reduced to permit the hydrolysed syrup to be offered at a very competitive price.

A model costing for production of hydrolysed whole whey is not at this stage available because of problems of confidentiality.

As far as permeate syrups are concerned these find application in brewing, soft drinks and in sugar confectionery where they may be regarded as alternatives to corn syrup. We have successfully produced beer containing 20% of its fermentables as permeate syrup.

A model costing for a hydrolysed lactose syrup produced from whey UF permeate is given in Table XII.

Table XII - Hydrolysed Permeate Cost Model

| | £/tonne solids | |
|---|---|---|
| Permeate Pretreatment | 30 | |
| Hydrolysis | 80 | |
| Demineralisation (50%) | 35 | |
| Evaporation | 120 | |
| Marketing | 10 | |
| | | 275 |
| Suggested Selling Price | | 300 |

Profit £25/tonne (0.14 pence/litre whey)

## Methane Production

Anaerobic fermentation of whey or permeate to produce methane which may be used to raise steam at the cheese factory is an attractive concept because it requires no marketing effort. A model costing is given in Table XIII.

Table XIII - Methane from Whey Cost Model

| | £/tonne whey |
|---|---|
| Operating Costs | 0.90 |
| Capital Costs | 0.68 |
| | 1.58 |
| Yield of gas = $35m^3$ value | 1.90 |
| 'Profit' | 0.32 |

Profit equivalent 0.03 pence/litre

## Lactitol

An interesting derivative of lactose is the polyhydric alcohol, lactitol, manufactured by the high pressure hydrogenation of lactose. It is only slowly fermented by oral bacteria and may therefore be regarded as "tooth safe". It is only partially metabolised and consequently has quite a low calorific value. As such it may take its place with other polyhydric alcohols such as sorbitol as a food ingredient. At this stage of development is is not possible to give any cost/benefit analysis.

## Summary

The following Table lists the benefits, in terms of current financial return per litre of whey for the various options I have considered. It should be noted that another measure, for example, return on capital invested, could give a significantly different picture.

Table XIV - Summary of Financial Return

| | pence/litre of whey |
|---|---|
| Methane Production | 0.03 |
| Lactose and Delactosed Whey Powder | 0.09 |
| Ultrafiltration Protein | 0.11 |
| Hydrolysed Permeate | 0.14 |
| Ion Exchange Protein | 0.29 |
| Whey Powder | 0.43 |

It should be noted that if permeate from ultrafiltration or ion exchange protein production is hydrolysed the profitability of the protein production and hydrolysed permeate should be added together.

Currently, a very conventional use for whey, the production of whey powder is quite profitable but the market for whey powder is quite volatile and therefore the current situation could change quickly.

Perhaps the conclusion to be drawn is that no one use for whey is so commercially advantageous that it will displace other uses.

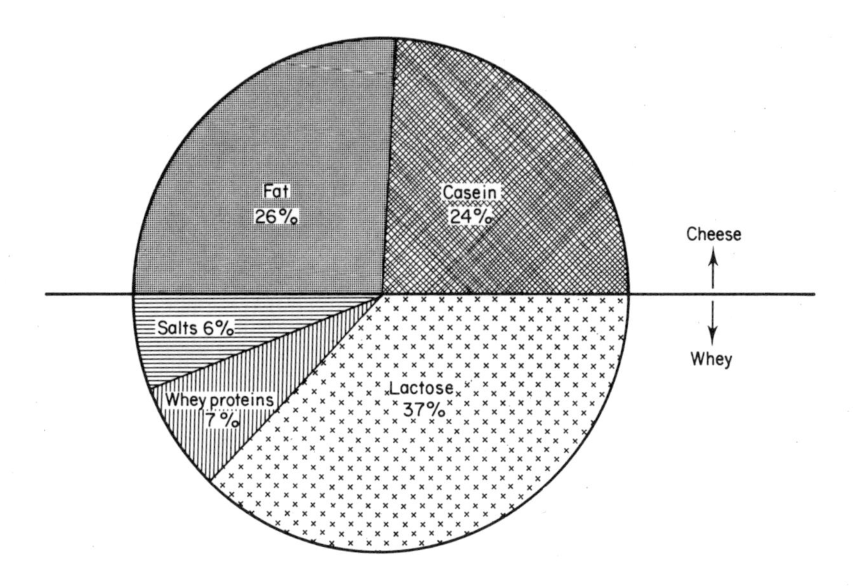


Figure 1 - Distribution of milk solids in cheese making

(After De Boer and Hiddink)

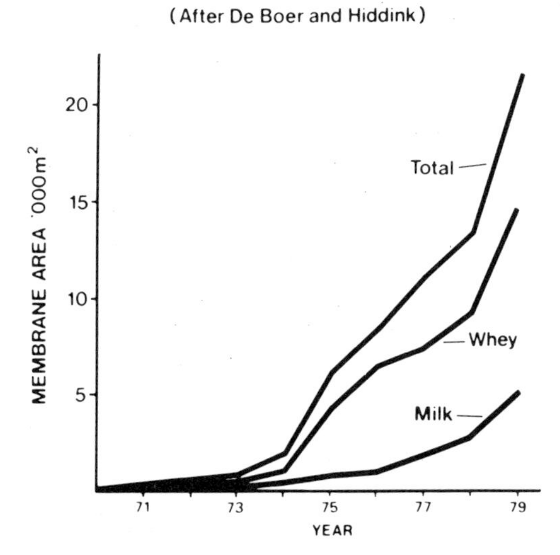


Figure 2 - World installed dairy ultrafiltration plant

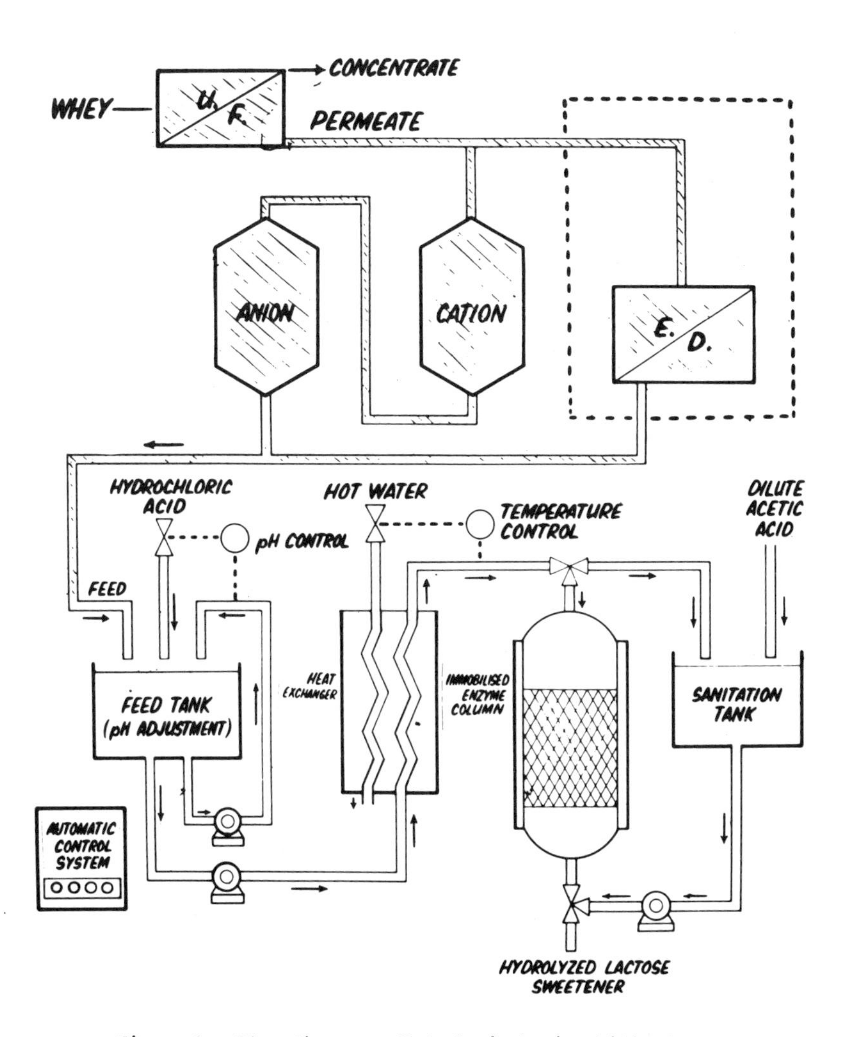


Figure 3 - Flow diagram - Hydrolysis by immobilised enzyme

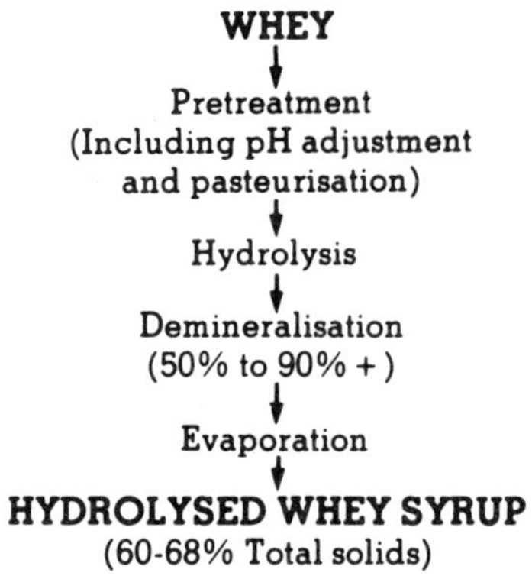

**WHEY**
↓
Pretreatment
(Including pH adjustment
and pasteurisation)
↓
Hydrolysis
↓
Demineralisation
(50% to 90% + )
↓
Evaporation
↓
**HYDROLYSED WHEY SYRUP**
(60-68% Total solids)

Figure 4 - Process sequences for the hydrolysis of lactose in whey

# UPGRADING OF MILK UF-PERMEATE BY YEAST FERMENTATION - SEMIINDUSTRIAL TRIALS AND ECONOMY

N. HALTER and Z. PUHAN
Laboratory of Dairy Science
O. KAEPPELI
Institute of Biotechnology
Swiss Federal Institute of Technology

## Summary

By introducing ultrafiltration in dairy technology, the problem of utilization of the permeate arises. Despite its water content of 95%, the permeate has a $BOD_5$ content of 30.000 to 45.000 mg $O_2$/l which does not allow direct disposal into public sewage.

The process of yeast fermentation of the permeate by using the strictly aerobic fodder yeast Trichosporon cutaneum offers the possibility of transforming it cheaply to animal feed. In this process, a simple technology is applied to transform part of the lactose under non-aseptic conditions into biomass at pH 4 and a temperature of 30$^{o}$C. The fermented permeate contains 0,5-0,8% protein, 1,3-2,0% lactose, 0,7-0,8% ash and 0,3-0,4% fat. The production costs and the feeding value are approx. equal.

## INTRODUCTION

While in processing of milk, it is important to manufacture high quality products for the successful utilization of milk, the use of raw material can only be considered optimal if the by-products of production are also being utilized.

According to Swiss milk statistics for 1981, 95% or 12,5 mio tons of whey are used for animal feeding, mainly directly to pigs, as this is the most economic solution because of the high water content of whey.

In many dairies, the technology of ultrafiltration (UF) has been introduced to concentrate milk or whey. The milk protein is thereby obtained quantitavely. UF is also used for obtaining whey proteins which are of high nutritional value. The permeate (filtrate) is a by-product of UF and, depending on the degree of concentration, it is obtained in considerable quantities. However, the main problem does not lie in the quantity but in the composition of the permeate. It contains 94-95% water, 4,5-5% lactose, 0,04% nitrogen components and approx. 0,5% ash. On the one hand, the economical use of the valuable constituents is influenced by the high water content. Because of these components the UF-permeate, having a $BOD_5$ content of 30.000 to 45.000 mg $O_2$/litre, cannot be directly disposed as waste water into sewage.

The process of yeast fermentation of permeate as described hereafter has been especially developed for the needs of the Swiss dairy industry. Fermentation is meant to be simple, economical and also applicable in smaller factories with a daily permeate volume of 5000 to 10.000 litres. As permeate is obtained daily, fermentation time should not exceed 12-14 hours and the protein content should reach at least 0,5-0,8%.

The marginal conditions for the process to be developed were as follows:

- Simple technology
- Fermentation time not exceeding one day
- Automatic process
- Low cost
- Fermented UF-permeate to be used directly as pig feed.

## ORGANISMS FOR FERMENTATION

Trichosporon cutaneum proved to be the most suitable yeast for the fermentation of permeate and was given preference to the fodder yeasts Saccharomyces lactis and Saccharomyces fragilis following comprehensive experiments. Trichosporon cutaneum metabolizes lactose in a strictly respiratory way to biomass, $CO_2$ and water, even at high concentration or limited oxygen supply. The yeast grows well at pH 4 which is at the same time a protection against bacterial contamination. Thus, the fermenter does not have to be asceptic and, therefore, requires a less sophisticated construction which means lower costs.

## FERMENTATION MEDIUM

For yeast fermentation, nutrients such as N-source, trace elements and vitamins must be added to the UF-permeate. The following additives are necessary for the complete utilization of lactose:

| | |
|---|---|
| - $NH_4Cl$ | 6,0 g/l permeate |
| - Trace elements | |
| . $CuSO_4 \cdot 5\ H_2O$ | 0,78 mg/l " |
| . $FeCl_3 \cdot 6\ H_2O$ | 4,8 mg/l " |
| . $ZnSO_4 \cdot 7\ H_2O$ | 3,0 mg/l " |
| . $MnSO_4 \cdot 2\ H_2O$ | 3,5 mg/l " |
| - Vitamines | |
| . Biotine | 0,01 mg/l " |
| . m-Inositale | 20,0 mg/l " |
| . Ca-Pantothenate | 10,0 mg/l " |
| . Pyridoxine-HCl | 0,5 mg/l " |
| . Thiamine-HCl | 2,0 mg/l " |

The first feeding trials showed that pigs liked the fully fermented permeate. From the nutritional point of view, however, problems arise because of the low dry matter content and the unfavourable C/N ratio.

## PARTIAL FERMENTATION OF LACTOSE

In order to alter the C/N ratio and increase the dry matter content at the same time, partial fermentation of lactose seemed to be a better solution. By limiting the N-source, the process was controlled in such a way that only approx. 50% of the lactose were utilized and transformed into biomass. Moreover, the reasons favouring this solution are less ash and reduction of costs.

Following partial fermentation, the protein content of the fermented permeate was similar to that of whey, i.e. 0,5-0,8%. Compared with full fermentation, the N-source had to be reduced by 50%, the vitamines and trace elements by 40%.

Experiments in a large Fermenter (5000 - 7000 litres)

After the results of experiments on a pilot scale proved satisfactory, a large fermenter was built to test the process under factory-like conditions and to obtain sufficient product for the feeding trials. The fermenter was installed in a cheese factory. A milk storage tank having a length of 2720 mm, a diameter of 2550 mm and a capacity of 12.000 litres was made into a fermenter. However, its dimensions were not optimal as a higher tank with a smaller diameter would have been preferable to increase the contact time of the oxygen with the medium. Moreover, the tank had to be installed horizontally due to the low ceiling in the factory. The working volume was 5000 to 7000 litres.

Aeration

In order to keep the construction as simple as possible, we installed a self-sucking stirrer (Ytron Y18, 5-2/T8-N) capable of distributing the air in fine bubbles in the medium. By the suction effect between stator and rotor with 3000 r.p.m., the air was drawn in through a by-pass tube. It was cleaned from dust particles with a cotton filter. In this way, 1,73 $m^3$ air/min. could be sucked into the medium. With a working volume of 5000 l, the aeration rate was 0,35 vvm. The mentioned stirrer did not allow better aeration.

Cooling

Temperature measurements revealed that a maximum of 12,5 kcal/h per litre of medium had to be removed. As a double-jacket or cooling with built-in tubes were too expensive, we mounted a ring of perforatet tubes on the outside of the tank. The surface of the fermenter was covered by a water film to obtain the nessessary cooling effect. Cooling was controlled by a temperature sensor in the fermenter as well as an on/off switch.

pH-control

At the beginning of the fermentation, the pH was adjusted to 4 with 33% HCl (1,9 ml/l). During fermentation, the medium was kept constant at pH 4 with NaOH (50%).

Foam formation

As the head space in the fermenter was large enough, it was sufficient to add antifoam P 2000 (0,5 ml/l) to limit foam formation. However, this antifoam amounts to 30% of the cost for medium supplements. In a fermenter for permanent production it would, therefore, be cheaper to instal a mechanical foam destroyer, thus avoiding the use of expensive antifoam agents.

Pasteurization of UF-permeate

Originally, it was intended not to pasteurize the permeate prior to yeast fermentation. Experience showed, however, that the danger of infection by wild yeasts was high, the reasons probably being that the permeate in the factory is regarded as waste product. Therefore, we decided to pasteurize the permeate prior to fermentation (72$^{o}$C, 15 sec.).

Culture preparation

The freeze-dried yeast culture was grown in a 15 litre laboratory fermenter. This pure culture was then scaled up in a 200 l fermenter under unsterile conditions to produce the starter culture for the 5000 l batch. The

200 l fermenter for the production of the mother culture was at the same time used as storage tank for the inoculum for the subsequent production. The culture was stored at 4°C.

Fermentation

The experiments on an industrial scale were carried out by fed-batch process. However, preliminarly laboratory experiments revealed that continuous fermentation was possible.

Fig. I: Process for upgrading of milk UF-permeate by fermentation with Trichosporon cutaneum

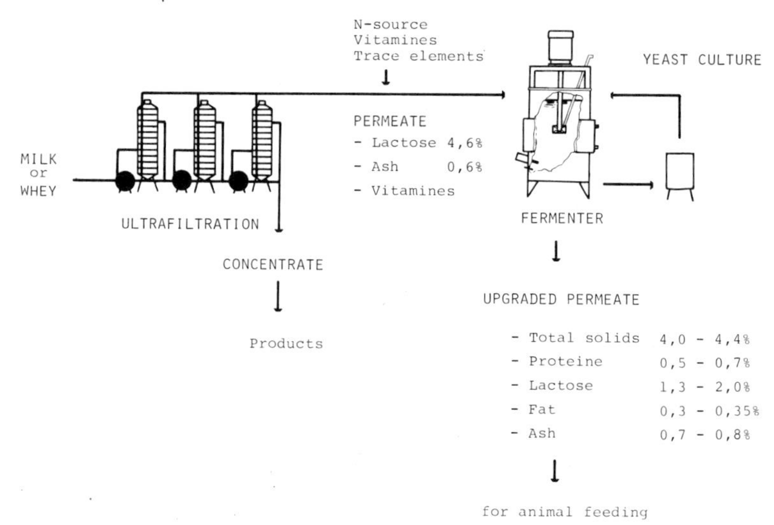


Fig. II: Growth of Trichosporon cutaneum in 3 subsequent batches of 5000 litres

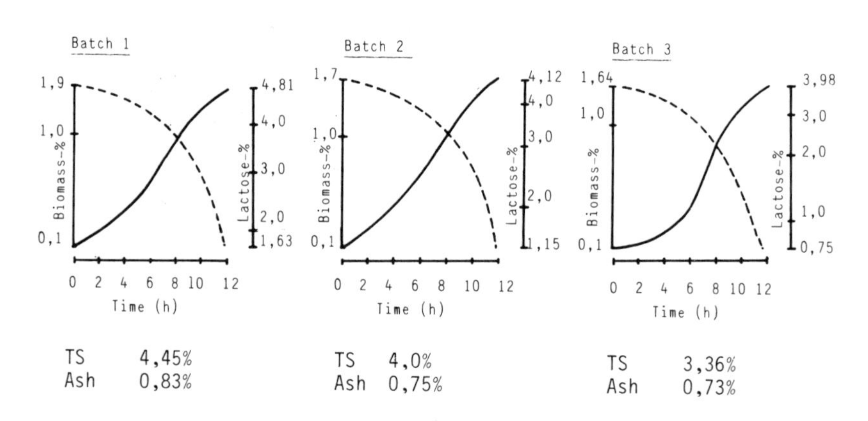

Composition of upgraded UF-permeate

| | |
|---|---|
| TS | 4,27% |
| Biomass | 2,00% |
| Lactose | 1,33% |
| Fat | 0,32% |
| Ash | 0,76% |
| Protein (Summ of aminoacids, average of 10 batches) | 0,71% |
| Galactose | 0,0034% |
| Glucose | 0,015% |

Aminoacides in protein - %

| | | | |
|---|---|---|---|
| Hydroxyproline | 0,0 | Methionine | 2,28 |
| Aspartic acid | 9,38 | Isoleucine | 4,2 |
| Threonine | 5,05 | Leucine | 7,85 |
| Serine | 5,16 | Tyrosine | 3,51 |
| Glutamic acid | 12,64 | Phenylalanine | 4,6 |
| Proline | 6,1 | Hydroxylysine | 0,0 |
| Glycine | 5,48 | Lysine | 8,1 |
| Alanine | 8,57 | Histidine | 2,56 |
| Cysteine | 0,93 | Arginine | 6,32 |
| Valine | 5,83 | | |

Minerals - %

| | |
|---|---|
| Sodium | 0,25 |
| Potassium | 0,11 |
| Calcium | 0,04 |
| Chloride | 0,44 |
| Phosphor | 0,04 |

Faty acids in fat - %

| | |
|---|---|
| Myristic | 0,8 |
| Palmitic | 22,1 |
| Palmitoleic | 0,4 |
| Heptadecenoic | 0,3 |
| Stearic | 18,9 |
| Oleic | 40,8 |
| Linoleic | 16,7 |
| Lactic acid | 0,11 |
| L(+) | 0,03 |
| D(-) | 0,077 |

ECONOMY

Assuming that yeast fermentation of UF-permeate is carried out in a dairy, the costs of equipment are estimated at sFr. 140.000.-. With a daily quantity of 10.000 litres of permeate and 250 working days per year, the cost for amortisation, payment of interest, maintenance and repair amounts to sFr. 0,01 per litre. The necessary medium supplements and energy costs for the UF-permeate are sFr. 0,022 - 0,034 per litre depending on the use uf antifoam. The total costs per litre for the fermentation of UF-permeate thus amount to sFr. 0,032 - 0,044 compared with the feeding-value of the fermented product which is sFr. 0,033 - 0,038 per litre.

In cost analysis it must be considered that in case of a daily disposal of 5000 - 10.000 litres of permeate into the sewage the costs would be considerably higher. According to Swiss regulations, the permeate would have to undergo biological treatment prior to leaving the factory.

## CONCLUSIONS

The partial fermentation process of UF-permeate with Trichosporon cutaneum offers the possibility of upgrading the permeate at low cost to pig feed with a feeding value based on dry solids nearly equal to that of whey. The process is simple, and because of the aerobic yeast Trichosporon cutaneum its control causes no problems. Production cost and average feeding value are approx. equal. If, however, the permeate has to be treated before being disposed in sewage, this process is economically feasible and offers a true alternative to the sewage disposal of UF-permeate.

# INDUSTRIAL PRODUCTION OF S.C.P. FROM WHEY

G. Moulin, B. Malige* and P. Galzy
Chaire de Génétique et Microbiologie - ENSA - INRA
Place Viala 34060 Montpellier Cedex - France

Summary

The nature of the balanced flora of an industrial fermenter is reported. The results show that it is an association of 3 species of yeast. The variation in yield is tied to the metabolism of the dominant species and especially to oxygen transfer in the fermenters. The variation of temperature, pH, dilution rate do not change flora balance. However, lactose concentration greatly affects yield for the bracket 20 g/l to 27 g/l. The balanced flora results thus in a great polyvalence in the treatment of whey and a great stability of production since only two dominant species will vary depending on the proportion of lactose and lactic acid.

## 1. INTRODUCTION

Whey, a by-product of cheese or casein manufactures, is produced in great quantities. In France this production corresponds to 450 000 metric tons of dry matter, of which 320 000 are lactose, and 60 000 are proteins. The proteins can be recovered by techniques recently developped (isoelectric point precipitation, ultrafiltration). In 1958, Fromageries Bel produced in France 800 metric tons food yeast per year from deproteinized whey. Since then the production increased ten-fold. From the beginning, the selection procedure has retained two strains belonging to the *Kluyveromyces* spp. In this paper we present the microbiological analysis of the flora in the fermentation towers after 20 years of operation. The influence of different culture parameters on the balance of the flora and on yield is also presented.

## 2. PRODUCTION CONDITIONS

The yeast plant where these experiments were performed has three 23 $m^3$ useful capacity air lift fermentation towers. The whole process is continuous according to the chemostat principle.

The wheys are deproteinized partly at high temperature (95°C) by a process of precipitation at protein isoelectric point (pH 4.6) and partly by an ultra-filtration process. Following protein separation, the wheys are mixed and stored at 20°C after being adjusted to pH 3 by addition of sulfuric acid. According to daily requirements, the whey solution is diluted so that its lactose concentration is between 20 and 25 g/l. Nitrogen addition is directly performed in the fermentation towers with solutions of ammonium sulfate and liquid ammonia. Fermentation pH (3.3) is regulated with a dilute solution of sulfuric acid. Following fermentation, the yeast is separated by centrifugation, washed, and concentrated by filtration. The yeast is then submitted to a thermo-mechanical cell plasmolysis process before flash sterilization at 130°C, drum drying, and packaging.

Physical and chemicals controls of the process, microbiological analyses and analytical methods have been previously described (1,2 ).

* C.R.V. Fromageries Bel 41100 Vendôme.

## 3. RESULTS

### 3.1 A balanced flora in the fermenters

A total of twelve sets of analysis were performed at the rate of twice weekly on samples from one of the three industrial fermentation towers. Table 1 shows the mean working characteristics of the fermenters. The results for the three dominant species show a very high proportion pf *Kluyveromyces fragilis* strains (90%). The *Torulopsis sphaerica* strains represented 9% of the total flora. The *Torulopsis bovina* strains never exceeded 1% of the total flora.

We chose several representatives from the isolated strains, of the different species in order to determine their growth and metabolic parameters under pure culture conditions. Lactose, lactic acid, ethanol, and glucose were used as carbon sources. These result have been published previously (1,2).

### 3.2 Hypothesis on the balanced flora

The isolation of the various strains and the determination of their parameters have prompted us to formulate the following hypothesis:

- The *Kluyveromyces fragilis* strains grow by using lactose. These strains show a strong Pasteur effect, which considerably reduces fermentation in the presence of air($Q^{air}_{CO_2}$ ferm=20 $\mu$l.h$^{-1}$.mg$^{-1}$). In order to verify this observation, we undertook the determination of the medium alcohol content evolution during a pure culture growth of *Kluyveromyces fragilis* in an aerated medium Ethanol was continuously formed at 0.06 mg ethanol h$^{-1}$.mg$^{-1}$ of D.W. Their performance on lactose,their share of the total population and the physiological properties (2) strengthen this hypothesis.
- The *Torulopsis bovina* strains grow on ethanol. These strains are characterized by their lack of fermentative metabolism ($Q^{N}_{CO_2}$=0). Their oxidative metabolism is strong in the presence of ethanol ($Q_{O_2}$=200) but rather weak on glucose ($Q_{O_2}$=50). These strains show a particularly well adapted metabolism for ethanol. The fact that they represent about 1% of the total flora is compatible with the quantities of ethanol produced by *Kluyveromyces fragilis* lactose utilization under the maximal aeration conditions of the plants.
- The *Torulopsis sphaerica* strains grow on lactic acid. Here again, the metabolism of these strains is best adapted to use lactic acid. Also the lactic acid content of whey, after dilution and treatment at the plant, is always in the order of 2 to 2.5 g/l. Under these conditions it is compatible with the hypothesis that *Torulopsis sphaerica* represent 10% of the total population. A few simple tests also suggest that the *Torulopsis sphaerica* population is higher in whey enriched with lactic acid.

### 3.3 The principal fermentation parameters' action on flora balance and yield

Many factors are involved in the operation of the fermenters. We tried to determine how the flora balance was affected when some parameters such as temperature, pH, dilution rate (D) and lactose concentration were modified independently or simultaneously.

Variation of temperature from 28° to 38°C or of pH from pH 2.8 to pH 4 does not affect the flora balance. The study of the influence of dilution rate shows that when D varied from 0.29 to 0.38 the proportion of the three species are not modified, but for D>0.33, a progressive decrease of biomass and lactose appearence are observed.

Lactose concentration does not affect the balance flora, but greatly affects yield (Table 2). During this study other parameters are maintained at the values of table 1. Residual lactose is less than 0.5 g.l$^{-1}$. However residual lactose increases for in-coming lactose content above 27 g/l. Thus under the conditions of table 1, the average production of the fermentation towers is about 100 kg of yeast per hour. It is, however, noteworthy that

Table 1. General characteristics of fermentation

| | | | |
|---|---|---|---|
| Useful volume | 1.22680 ± 200 | Air flow $m^3.h^{-1}$ | 1800 ± 100 |
| Feed rate $l.h^{-1}$ | 7600 ± 150 | Cell concentration $g.l^{-1}$ | 13.3 σ 0.4 |
| Dilution rate $h^{-1}$ | 0.33 ± 0.01 | Residual lactose $g.l^{-1}$ | 0.4 σ 0.2 |
| pH | 3.0 ± 0.05 | Dry matter yield % | 56.0 σ 2.2 |
| Temperature °C | 38 ± 1 | Productivity $g.l^{-1}h^{-1}$ | 4.4 σ 0.1 |
| Lactose $g.l^{-1}$ | 24 ± 1.5 | Protein content of yeast % | 55 σ 0.7 |

Table 2. Yields at variable lactose concentrations

| Lactose $g.l^{-1}$ | Pumping rate $l.h^{-1}$ | Biomass $g.l^{-1}$ | Yield (Biomass vs lactose) | Available lactose $kg/h^{-1}$ | Yeast Production $kg/h^{-1}$ |
|---|---|---|---|---|---|
| 20.20 | 7 600 | 12.17 | 61.05 | 152.54 | 92.49 |
| 21.87 | 7 667 | 13.41 | 65.23 | 157.61 | 102 81 |
| 21.35 | 7 562 | 13.33 | 63.23 | 159.40 | 100.8 |
| 21.85 | 7 668 | 13.20 | 61.26 | 167.54 | 101.2 |
| 21.90 | 7 584 | 12.65 | 58.47 | 164.07 | 95.9 |
| 22.10 | 7 566 | 13.25 | 60.70 | 165.15 | 103.2 |
| 22.60 | 7 587 | 12.55 | 56.18 | 169.48 | 95.2 |
| 23.85 | 7 598 | 12.86 | 54.65 | 178.78 | 97.7 |
| 24.35 | 7 744 | 12.44 | 51.68 | 186.41 | 96.3 |
| 24.72 | 7 617 | 13.00 | 53.32 | 185.71 | 99 |
| 24.73 | 7 528 | 12.48 | 51.15 | 183.67 | 93.9 |
| 26.10 | 7 650 | 12.23 | 47.47 | 197.06 | 93 |

this production requires 152 to 197 kg of lactose per hour. This variation in yield is tied to the metabolism of the dominant species (Kluyveromyces fragilis) and especially to oxygen transfer in the fermenters. In all cases, dry matter content of the fermenter is between 12 and 13 g/l, and in the presence of high concentration of lactose, excess lactose is transformed into ethanol as there is lack of oxygen. According to our hypothesis, ethanol must be utilized by Torulopsis bovina strains. It is probable that under the fermenters' present operation conditions (lactose 24 g/l) the limited oxygen availability does not permit the oxydation of the entire ethanol formed.

## CONCLUSION

Yeast production on whey was considered as a pure culture. The above results show that actually there is an association of 3 species. The study of metabolic parameters of the strains as pure cultures, and the observations collected during 2 years of plant operation allowed us to formulate a working hypothesis on this balanced flora.

The balanced flora results thus in a great polyvalence in the treatment of whey and a great stability of production since only the relationship betwen the two dominant species will vary depending on the proportions of lactose, lactic acid. All the carbon substrates present or formed during the process is utilized.

## REFERENCES

1. MOULIN, G., MALIGE, B. and GALZY, P. (1983). Balanced flora of an industrial fermenter: Production of yeast from whey. J. Dairy Sci. 66: 21.
2. MOULIN, G., MALIGE, B. and GALZY P. (1981). Etude physiologique de Kluyveromyces fragilis: conséquence sur la production de levure sur lactosérum. Le Lait, 61: 323.

# STUDY OF S.C.P. PRODUCTION FROM STARCH

G. Moulin, F. Deschamps* and P. Galzy
Chaire de Génétique et Microbiologie - ENSA - INRA
Place Viala 34060 Montpellier Cedex - France
*I.R.C.H.A. - Centre de Recherche 91700 Vert-le-Petit - France

## Summary

Fifteen yeast strains were selected for the production of food yeast from starchy substrates. From the comparison with the amylolytic yeasts, a strain of Schwaniomyces castellii was selected and its characteristics are described.

## 1. INTRODUCTION

The production of food yeast or protein enriched starch may be possible solutions. As early as 1944, Wickerham et al (1) proposed the use of a yeast, Endomycopsis fibuligera Lindner-Decker, for the production of single cells from waste waters of potato processing plants.

Many authors attempted to select strains for the production of protein, Hattori (2) selected an Endomyces sp. among 27 strains belonging to several species. Spencer Martin and Van Uden (3) recommended a Lipomyces kononenkoe strain among 81 strains from different species. In both cases the authors underlined the amylase exocellular character of the strains selected. We showed (4) that for most amylolytic strains, the excretion of amylases depends on the composition of their growth medium. Others authors reported the excretion of amylases by Schwaniomyces alluvius (5), Endomycopsis fibuligera (6), Lipomyces kononenkoe (7). In these cases, the media used were favorable for the enzymes excretion. There is also a close relationship between excretion of amylases and dry-matter yield. Until now our selection procedures often used a medium (Yeast Nitrogen base difco) unfavorable to the excretion of amylases. We were thus driven to repeat our strain selection procedure using four other different substrates. We will also review in the present paper the parameters used in the assesment of some strains recommended for the production of SCP from starch.

Experimental conditions have been previously described.

## 2. RESULTS

### 2.1 Strains selection

Generation time, residual starch and final pH were measured for each strain. Some strain did not hydrolyse starch at all or very little, under any condition. On the other hand, the starch was completely hydrolyzed by 7 strains when the media conditions allowed the excretion of enzyme as it has been described (4). Lipomyces starkeyi hydrolyzed starch completely, however its generation time was very long. Pichia burtonii seemed to be very sensitive to pH variations. The other 5 strains: E. fibuligera, S.castellii, S. alluvius, S. occidentalis and T. ingeniosa showed similar generation times which were compatible with industrial condition requirements.

### 2.2 Growth study of Schwaniomyces castellii

This strain was further tested on three media: Starch Y.N.B., Starch Y.N.B.-buffer and cassava flour medium with pH adjusted to 3.5. The results are presented in table 1. The amount of starch consummed was very much improved by buffering the media. This observation had to be tied to the excretion of amylase which was a lot of more important with this last two-media as compared to starch-Y.N.B. medium.

The main difference in the results obtained with growth on buffered starch Y.N.B. as compared to cassava flour medium was the protein content of the dry-matter produced. In the case of cassava flour medium about 15 % of the initial medium dry-matter were not attacked and were separated out by centrifugation. This non-assimilated material was neither hydrolyzed by amylases nor by chemical means. This phenomenon was responsible for the high dry-matter yield shown in table I . The protein content of the corresponding final dry-matter was decreased probably due to the addition of yeast extract 1 g/l. The continuous growth tests were performed using the cassava flour medium at 35°C, pH 3.5. The yield results presented in table I show that it was possible to use up entirely the hydrolyzable starch with a dilution rate of 0.23 $h^{-1}$.

## 3. DISCUSSION

The general characters of somes strains and processes studied by differents authors are summarized in table II.

Two processes applied a starch pre-hydrolysis step. Only the Symba process was developped to industrial scale (8). This process is characterized by a two step technology, working under sterile condition with a relatively low dilution rate. Moreton (9) also recommended the C. utilis strain for chemically or enzymatically prehydrolyzed starch. Under these conditions, this strain growth rate was very high. On the other-hand the cost of the pre-hydrolysis step of the process was an important draw-back.

The other three schemes proposed the use of a pure single culture in a one step process. Important features of the strains utilized are also shown in table II. S. castellii and L. kononenkoe excrete their amylases into the media. In the case of C. tropicalis, the authors (10) underlined the fact that the enzyme was not excreted. This was probably due to the medium characteristics. S. castellii has a higher growth temperature (35°C) and the ability to multiply at a lower pH (3.5) than the other strains. These two features are helpful in an industrial production operation. The estimated yields for each process are similar.The growth rates of C. tropicalis and S. castellii are close to those usually reported for an industrial production of yeasts on other substrates. However, strain L. kononenkoe shows a relatively slow growth rate. The continuous culture experiments performed by the different authors can not be compared with each other, considering the differences in experimental conditions. The main parameters confirmed however, the results in batch culture. In particular optimum dilution rate was 0.26 $h^{-1}$ for C. tropicalis 0.23 $h^{-1}$ for S. castellii and only 0.12 $h^{-1}$ for L. kononenkoe.

| | $\mu$ | Protein content of final product (%) | D.M. yield as % of initial starch used | D.W. yield as % of assimilated starch |
|---|---|---|---|---|
| Starch-YNB | 0.19 | 42 | 26 | 45 |
| Starch-YNB pH 5.5 | 0.23 | 41 | 40 | 44 |
| Cassava flour pH 3.5 | 0.23 | 34 | 44 | 48 |

TABLE I - The main parameters values for Schwaniomyces castellii on 3 media

| Strain | Amylase type | Excretion | T°C | pH | Fermentation type | Yield assimilated sugar | μ |
|---|---|---|---|---|---|---|---|
| E.fibuligera(8) C. utilis | α-amylase amyloglucosidase | Strong | 30° | 4.5 | Continuous two-steps | 0.43 | 0.10 |
| C.tropicalis(10) | Undeterminet | Nil | 32° | 4 to 5.5 | Batch one step | 0.45 to 0.55 | 0.25 to 0.45 |
| L.kononenkoe(11) | α-amylase amyloglucosidase | Strong | 28° | 5.5 | Batch one step | 0.5 | 0.14 |
| S. castellii | α-amylase amyloglucosidase | Strong | 35° | 3.5 | Batch one step | 0.45 | 0.28 |
| C. utilis (9) | no-amylase pre-hydrolyzed starch | - | 30° | 4.0 | Batch one step | 0.54 | 0.5 |

TABLE II- Main features of the recommended strains for the production of S.C.P. from starch.

REFERENCES

1. Wickerham L.J., Lockwood L.B., Petitjohn O.G. and Ward G.E. 1944. Starch hydrolysis and fermentation by the yeast Endomycopsis fibuligera. J. Bacteriol. 48, 413-419.
2. Hattori Y. 1961. Studies on amylolytic enzymes produced by Endomyces sp. Part I. Production of extracellular amylase by Endomyces sp. Agr. Biol. Chem. 25, 737-743.
3. Spencer-Martins and Van Uden N. 1977. Yields of yeast grown on starch European J. Appl. Microbiol. 4, 29-35.
4. Oteng-Gyang K., Moulin G. and Galzy P. 1980. Influence of amylase excretion on biomass production by amylolytic yeasts. Acta Mikrobiol. Hung. 27, 155-159.
5. Calleja G.B., Moranelli F., Nasim A., Duck P.D. and Levy S. 1981. Monitoring population growth and extracellular amylolytic activity in Schwaniomyces species. VIIth Int. Specialized Symposium on yeast. Valencia - Spain, September 1981.
6. Clement F., Rossi J., Costamagna L. and Rosi J. 1980. Production of amylase by Schwaniomyces castelli and Endomycopsis fibuligera. Anton. van Leeuvenhoek, 46, 399-405.
7. Spencer-Martins and Van Uden N. 1979. Extracellular amylolytic system of the yeast Lipomyces kononenkoae. Eur. J. Appl. Microbiol. 6, 241-250.
8. Jarl K. 1969. Symba yeast process. Food. Technol. 23, 23-26.
9. Moreton R.S. 1978. Growth of Candida utilis on enzymatically hydrolysed potato waste. J. of Appl. Microbiol. 44, 373-382.
10. Azoulay E., Jouanneau F., Bertrand J.C., Raphael A., Jansens J. and J.M. Lebeault. 1980. Fermentation methods for protein enrichment of cassava and corn with Candida tropicalis. Appl. Anviron. Microbiol. 39, 41-47.
11. Correia Sa and Van Uden N. 1981. Production of biomass and amylases by the yeast Lipomyces kononenkoae in starch limited continuous culture. Eur. J. Appl. Microbiol. Technol. 13, 24-28.

# WHEY AS A SOURCE FOR MICROORGANISMS/AMINO ACID PATTERN

T. YAZICIOĞLU
Department for Nutrition and Food Technology,
TÜBİTAK, Marmara Research Institute-Gebze, TURKEY

Summary

Since whey is not utilized properly in Turkey at present, it could be a suitable raw material for SCP production. Estimated whey to be utilized in Turkey amounts to 500.000 tons.
For our trials, we have used some 22 different microorganisms; 12 of them were yeasts and the rest were moulds. In our trials the yields varied from 0.148 to 1.093 g protein in biomass, from 100 ml of medium. So it looks rather possible to produce 8.000-15.000 tons of biomass, or 3.000-5.000 tons of protein, in dry basis, from 500.000 tons of whey in Turkey. This amount of protein corresponds to the protein content of 30.000-50.000 t. of barley.
We have determined the amino acid compositions of the SCP produced in our laboratory trials and have compared the results with the FAO-provisional amino acid pattern. The contents of all essential amino acids of some SCP produced in our laboratory were higher than those of FAO-pattern. Some of our experimental SCP had rather high lysine contents. Such SCP could be used to fortify wheat flour, especially in developing countries, which consume much cereal products.

## 1. INTRODUCTION

According to the opinion of the experts, the classical foods will not be sufficient in the future to meet the nutritional needs of the growing world population. So it will be necessary to find some novel foods especially rich in protein. It is believed that the SCP could be one of them.

For trials of SCP production in Turkey, we have preferred to use first agroindustrial wastes, such as whey, black water of olive and vinasse of distilling industries. They are usually wasted in Turkey at present and they cause environmental problems (1, 2, 3).

We made recently in our Department some trials on production of lipid yielding microorganisms and we obtained biomasses containing 20-25 % of lipids and 12-21 % of protein (4).

We produce about 230.000 tons of cheese yearly and the estimated residual whey, mostly after having made whey cheese, is about 500.000 tons.

## 2. MATERIALS AND METHODS

We have used whey as nutrient media. We have obtained it from the neighborhood of Gebze, Sea of Marmara Region.

We prepared the inoculum as follows: We poured, in 250 ml erlenmeyer flasks, 50 ml of whey at pH 5.5 and sterilized, inoculated with desired microorganism and shaked 72 hours at a temperature of 30$^{o}$C for yeasts, and 28$^{o}$C for moulds (1). After that these suspensions were used to inoculate the whey in the fermenter. We have used a 5 liters fermenter with automatic temperature, air, foam and RPM control. 0.5 % Ammonium sulfate and 0.1 % monopatassium phosphate were added as nutrients into the whey. After sterilization, the whey in the fermenter, is inoculated with the inoculum.

The fermentation lasted usually for 72 hours. After fermentation, the relevant microorganisms is separated by centrifuging, dried in vacuum and weighed. We made our trials with 22 different microorganisms. 12 of them were yeasts (Candida lipolytica, C. robusta, C. utilis, Geotrichum candidum, Kluyveromyces lactis, Rhodotorula glutinis, R.pilimanae, R. gracilis, Torulopsis magnoliae, Torula utilis, S. carlsbergensis, S. cerevisiae) and 10 of them were moulds (Aspergillus niger, A. oryzae, A. ustus, Fusarium spec; Fusarium moniliformae, Mucor pusillus, Penicillium chrysogenum, P. notatum, P. oxalicum, Rhizopus spec.).

The amino acid compositions of the experimental SCP were determined in Multichrom Liquid Column Chromatograph No 4255 of the firm Beckman.

## 3. RESULTS

### 3.1. Composition of whey

The composition of Turkish whey we have used in our laboratory trials is given in Table I. pH-value of the whey was 5.5 (1).

TABLE I

| | |
|---|---|
| Water, % | 92 |
| Lactose, % | 5 |
| Protein, % | 0.85 |
| Fat, % | 0.3 |
| Ash, % | 0.3 |
| Riboflavin, mg/100 ml | 0.10 |
| Niacin, mg/100 ml | 0.10 |
| Iron, mg/100 ml | 0.12 |
| Sodium, mg/100 ml | 14.5 |
| Potassium, mg/100 ml | 30.0 |
| Calcium, mg/100 ml | 13.5 |

### 3.2. Yields

At the end of the fermantation process yields as g/100 mg, percentages of protein and g protein in biomass for each of the SCP were determined.

In Table II, the results of microorganisms with better yield and higher protein content are given. This table is arranged according to the yield of protein from 100 ml of medium.

So it is possible to produce 8.000-15.000 tons of biomass, or 3.000-5.000 tons of protein in dry basis from 500.000 tons of whey in Turkey. This amount of protein corresponds to the protein of 30.000-50.000 tons of barley.

### 3.3. Amino acid compositions

In order to have an Idea about the biological value of SCP produced in our laboratory from different kinds of microorganisms, we have determined their amino acid compositions, as well. In Table III only their essential amino acid contents are given and for comparison the FAO-reference pattern is also added (5).

TABLE II

| Microorganisms | Yield (g/100 ml medium dry basis) | Protein, % (in biomass, in dry basis) | g protein in biomass, from 100 ml |
|---|---|---|---|
| Rhizopus | 2.9 | 37.7 | 1.093 |
| Fusarium | 2.9 | 28.9 | 0.838 |
| Candida robusta | 1.6 | 41.0 | 0.656 |
| Penicillium oxalicum | 2.1 | 27.5 | 0.578 |
| Torula utilis | 2.2 | 26.1 | 0.561 |
| Aspergillus niger | 1.9 | 26.4 | 0.509 |
| Mucor pusillus | 2.0 | 24.9 | 0.503 |
| Mucor chrysogenum | 1.7 | 28.2 | 479 |
| Fusarium moniliformae | 1.1 | 43.3 | 476 |
| Rhodotorula pilimanae | 1.5 | 31.7 | 456 |

TABLE III

Essential Amino Acid Patterns of Protein of Biomass from Whey and for Comparison FAO-Pattern (Grams of Amino Acid per 100 g of protein)

| | Isoleucine | Leucine | Lysine | Phenylalanine | Tyrosine | Methionin and Cystine | Threonine | Valin |
|---|---|---|---|---|---|---|---|---|
| Aspergillus niger $M_1$ | 2.6 | 3.9 | 3.2 | 2.0 | 1.9 | 0.6 | 2.1 | 1.9 |
| Aspergillus orzyzae, As 31 | 4.8 | 7.2 | 4.4 | 2.9 | 3.0 | 3.4 | 4.1 | 0.5 |
| Fusarium | 8.9 | 9.3 | 5.7 | 4.2 | 3.2 | 1.8 | 5.1 | 7.4 |
| F. moniliformae | 2.8 | 4.2 | 3.2 | 2.1 | 2.8 | 0.4 | 2.3 | 3.4 |
| M. pusillus | 4.1 | 5.8 | 5.5 | 2.7 | 3.6 | Trace | 3.3 | 4.2 |
| P. chrysogenum | 4.7 | 6.7 | 5.0 | 2.4 | 3.2 | 1.8 | 3.7 | 5.1 |
| P. notatum | 4.8 | 5.7 | 7.8 | 5.2 | 6.5 | 3.2 | 3.7 | Trace |
| P. oxalycum | 3.9 | 5.2 | 20.7 | 3.4 | 4.5 | 22.3 | 4.7 | 3.0 |
| Rhizopus | 2.9 | 7.4 | 7.6 | 3.0 | 8.2 | Trace | 4.2 | 9.7 |
| C. lipolytica | 4.1 | 6.4 | 2.9 | 3.4 | 4.6 | 1.3 | 3.9 | 5.3 |
| C. robusta | 4.2 | 5.4 | 4.8 | 3.0 | 2.3 | 1.0 | 2.8 | Trace |
| G. candidum | 1.7 | 3.7 | 3.9 | 2.7 | 1.9 | 1.2 | 4.9 | 3.4 |
| K. lactis | 0.7 | 1.1 | 1.1 | 0.5 | 0.5 | 0.2 | 0.5 | 0.9 |
| R. pilimanae | 4.0 | 6.3 | 4.5 | 3.7 | 2.2 | Trace | 3.8 | 4.7 |
| S. carlsbergensis | 4.0 | 6.2 | 5.4 | 3.6 | 4.2 | 1.7 | 3.2 | 4.9 |
| S. cerevisiae | 4.5 | 6.7 | 5.4 | 2.8 | 0.9 | Trace | 3.9 | 4.8 |
| T. magnoliae | 5.7 | 7.5 | 7.1 | 3.5 | 2.6 | Trace | 5.0 | 6.4 |
| T. utilis | 5.6 | 7.6 | 7.7 | 4.7 | 6.3 | 6.4 | 5.6 | 6.4 |
| B megaterium | 4.8 | 5.7 | 5.5 | 3.9 | 1.3 | 4.1 | 3.6 | Trace |
| FAO-Reference Pattern | 4.2 | 4.8 | 4.2 | 2.8 | 2.8 | 4.2 | 2.8 | 4.2 |

We have noticed with a great interest, that each of the essential amino acids of Torula utilis are quantitavely higher than those of the FAO-pattern. The high lysine content of some SCP, such as especially of P. oxalicum and C. lipolitica is rather striking. Since lysine is a limiting amino acid of wheat protein, it could be important for developing countries, consuming much cereal products, to enrich them with those SCP, rich in lysine, in the future.

## REFERENCES


1. ÖCAL, Ş.-ARAN, N.-ÇELİKKOL, E. (1977). "Production of Microbial Protein from Black Water of Olive and Whey", Published by Marmara Research Institute, No. 26, Gebze, Turkey.
2. YAZICIOĞLU, T. and COWORKERS. (1980). "Some Trials of The Utilization of Whey, Black Water of Olive and Vinasse For Production of SCP in Turkey", Published by Marmara Research Institute, No. 42, Gebze, Turkey.
3. ÖMEROĞLU, S. (1978). "The Amino Acid Compositions of SCP obtained in Our Laboratory", Published by Marmara Research Institute, No. 28, Gebze, Turkey.
4. ARAN, N. (1981). "Some Trials on Production of Lipid Yielding Microorganisms", Published by Marmara Research Institute, No. 53, Gebze, Turkey.
5. ANONYMUS. (1965). "Protein Requirements", Published jointly by FAO and WHO. FAO Nutrition Report Services, No. 37, Rome.

# SCP PRODUCTION FROM WHEY: SCALE-UP OF A PROCESS

M. MORESI
Istituto di Chimica Applicata e Industriale
Università degli Studi di Roma
Italy

Summary

The methodology used in the development of a whey fermentation by *Kluyveromyces fragilis IMAT 1872* and reviewed here involved experimentation on shake -flask, laboratory- and pilot-fermenters and a statistical analysis of the results, thus obtaining: (1) an efficient screening of the operating variables to be further studied in the transference from the smaller scale to the larger one; (2) optimization of the biomass yield; (3) a scale-up criterion based on constant volumetric oxygen transfer coefficient (measured by the sulphite method) and constant air velocity. On the basis of these results and the data reported in the literature it was possible to outline the flow diagram of a yeast-whey protein production process and to estimate the processing costs by varying the plant size and substrate cost.

## 1. INTRODUCTION

Whey is the major byproduct of cheese manufacture and *ca.* 26% of the about 20 million $m^3$ of whey annually produced in the EEC countries are disposed of without any treatment, thus causing a tremendous pollution problem (the $BOD_5$ of raw whey ranges from 30,000 to 50,000 ppm).

Although the fermentation of whey by various microorganisms has been known and studied for years, the Ministry of Public Instruction of Italy has financed a research project aimed at assessing the techno-economic feasibility of SCP production from whey.

The main aim of this communication is to review either the scale-up procedure applied to examine such a process, or the main results of the scale-up exercise concerned.

## 2. SCALE-UP PROCEDURE AND RESULTS

Since the development of a fermentation process is an on-going exercise and requires a continuous assessment of its profitability, it was found to be practical to scale-up the process in several stages and submit the results to statistical analysis to limit the number of variables under study in the transference of the process from the smaller scale to the larger one. In this way, it was possible to investigate a large number of experimental variables in a short time with low expenditure.

2.1 Experimentation in the Shaken-Flask Scale(50 cm$^3$)

After a microbial screening the biomass yield y of the strain selected(*Kluyveromyces fragilis IMAT 1872 - Perugia, Italy*) was assessed by varying several operating variables, such as temperature(T), pH, medium composition(initial lactose L, ammonium sulphate AS, potassium posphate PP, and yeast extract YE concentrations), and fractional filling of the shaken-flasks(R). Factor analysis was firstly used to determine the controlling parameters of this process(*viz.* L and R)( 1). Then, by using the lowest value of R(0.05) a new series of experiments designed according to the *composite design* technique allowed not only the optimal conditions for y to be determined(T=36.4°C; pH=5.1; AS=PP=0.47%w/v; YE=0.11%w/v; L=24.75 g.dm$^{-3}$; y=0.432 g dried cells per g initial lactose), but also the greater influence of L and T than pH, salts and YE on y to be assessed( 2).

2.2 Experimentation in the Laboratory-Fermenter Scale(10 dm$^3$)

In this scale the optimal conditions given above were checked at different values of L, T, aeration(A) and stirring rate(N)( 3). Fig. 1 shows a series of loci at constant values of y and COD reduction efficiency $\eta$ as functions of L and N at constant T and A: the dashed region shows the values of L and N associated with a COD removal greater than 90% and a biomass yield ranging from 50 to 56%( 4).

2.3 Experimentation in the Pilot-Fermenter Scale(80 dm$^3$)

In this scale the effects of L and N on y and $\eta$ were confirmed by basing the scale-up on constant volumetric oxygen transfer coefficient $k_La$ and constant air velocity(5,6). By using the sulphite method to compare the oxygen transfer capability of the laboratory- and pilot-fermenters used, it was possible to develop the following empirical model( 6):

$$y=57.23-0.03(L-15.3)^2-1.07\times10^{-5}\times(k_La-1667)^2$$

$$\eta=91.29-3.56\times10^{-5}(k_La-1176)^2$$

where L is expressed as g.dm$^{-3}$, $k_La$ as h$^{-1}$, and y and $\eta$ as *percent* of initial lactose and COD, respectively. The validity of the above model was then checked against the biomass yields obtained in a 6-m$^3$ fermenter(6), using different types of whey, and directly fermenting deproteinized whey to obtain a yeast-whey protein product(YWP) of higher biological value than yeast alone.

Fig. 2 shows a typical fermentation run in the pilot-fermenter scale (5).

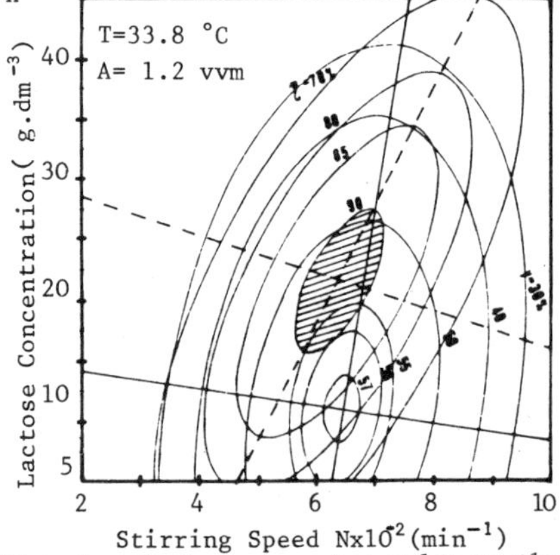


Fig. 1 - *Contour map of y and $\eta$ vs. the operational variables L and N at constant values of T and A as extracted from (3,4).*

### 2.4 Techno-Economic Analysis of a YWP Production Process

Details of the flow diagram and cost estimates for a YWP production process were reported in (7-9): *e.g.* referring to a zero substrate cost and a plant capacity of 10,000 tons/yr such a product($0.44/kg) would cost more than soybean($0.25/kg). Nevertheless, owing to its high protein score further scale-up of the process to the pilot-plant scale will be carried out to minimize SCP production costs.

### REFERENCES


1. MORESI, M., NACCA, C., NARDI, R. and PALLESCHI, C. (1979). Eur. J. Appl. Microbiol. Biotechnol. *8*:49-61
2. MORESI, M. and SEBASTIANI, E. (1979). *ibidem 8*:63-71
3. MORESI, M., COLICCHIO, A. and SANSOVINI, F. (1980). *ibidem 9*:173-183
4. MORESI, M., COLICCHIO, A., SANSOVINI, F. and SEBASTIANI, E. (1980). *ibidem 9*:261-274
5. BLAKEBROUGH, N. and MORESI, M. (1981). *ibidem 12*:173-178
6. BLAKEBROUGH, N. and MORESI, M. (1981). *ibidem 13*:1-9
7. MORESI, M. and SEBASTIANI, E. (1981). Studies in Environm. Sci. *9*:37-53
8. MORESI, M. (1981). Chim. e Ind.(Milan) *63*:593-603
9. MORESI, M. (1981). *ibidem 63*:731-739


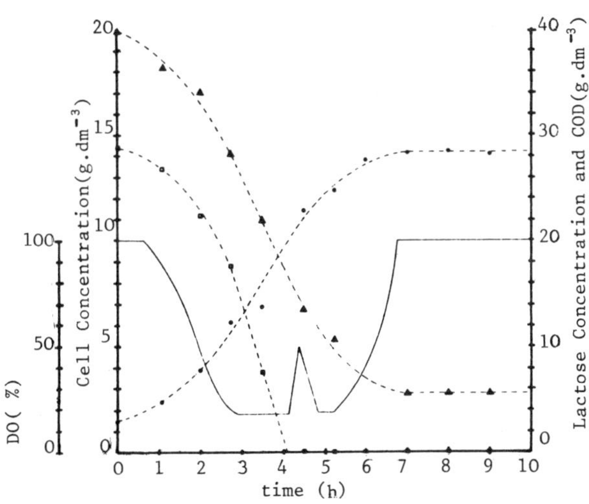


Fig. 2 - *Time course of a whey fermentation by* K. fragilis IMAT 1872 *at high lactose(28.86 $g.dm^{-3}$) and stirrer speed(450 $min^{-1}$) levels and T=35°C in a 100-$dm^3$ pilot-fermenter as derived from ( 5): ●, cell concentration; □, lactose concentration; ▲, COD. The continuous line refers to the dissolved oxygen(DO) concentration in the culture medium.*

## SUBJECT AREA 3

## NUTRITION AND TOXICOLOGY

Chairman : Ch. SCHLOTTER

# THE ANIMAL NUTRITIONISTS DREAM OF A NEW SCP

C. WENK
Institut for Animal Production, Nutrition Group, ETH-Zentrum, CH-Zurich

## 1. INTRODUCTION

In the last years new Single Cell Protein products (SCP) have been introduced as feedstuffs in animal nutrition mainly as protein sources of a high technological standard. On the other hand the nutritional improvement of waste products of various origines by microbial fermentation is in discussion and partly in use. Such feedstuffs like all others have to contribute to the supply of nutrients of the animals. The coherence of the nutrient content of the feedstuffs and the needs of the animals is shown in figure 1. Furthermore the feedstuffs should not contain nutrients of other substances in amounts, which can have a negative effect on the animals performance or which are evan toxic.

Figure 1: Coherence of the nutrient content of the feed and the nutrient needs of the animals

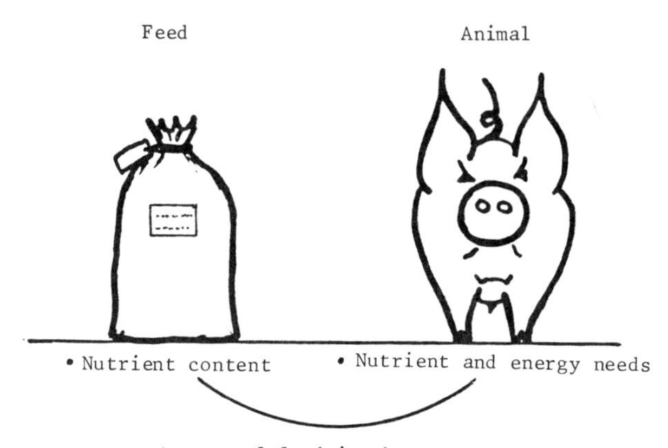


The animal covers its needs of nutrients and energy with the ingested feed. Besides the nutrient content of the feed, which can be composed of several feedstuffs, the amount of intake and also the utilization of the nutrients in the digestive tract (digestion) and in the tissues determine the anount of feed to be supplied.

## 2. NUTRIENT CONTENT OF THE FEED AND THE NUTRIENT NEEDS OF THE ANIMALS

In table 1 the relationship of the nutrient content of the feed and the needs of the animal is shown.

Table 1: How the animal covers its needs

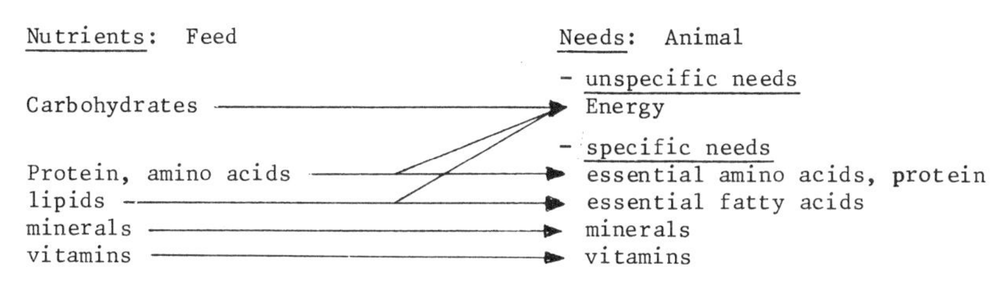


The needs of the animals have to be divided into the specific and the non specific needs. The specific needs comprise about 50 essential nutrients like amino acids, fatty acids, minerals and vitamins. These nutrients must be ingested with the feed, because they are essential for life and because they cannot be synthesiezed in the tissues. Amino acids are used by the animal in the L-form.

The animals specific needs depend on the species, age, weight and mainly the form and quantity of production like meat,milk, eggs or wool. A supply of the specific nutrients below the needs causes reduced performances and even specific deficiency symptoms. An intake moderately above the needs has usually no effect on the animals health. A certain surplus in essential nutrients is normal in animal nutrition.

A supply of specific nutrients far above the needs can cause an undesired reduction of feed intake, diarrhoea or even intoxication of the animals. In Table 2 some examples of the maximum tolerance level for dietary minerals are presented for different species of domestic animals (NRC, 1980).

Table 2: Maximum tolerable levels of dietary minerals for domestic animals

| Element | | Species<br>Cattle | Sheep | Swine | Poultry | Horse |
|---|---|---|---|---|---|---|
| Aluminium | ppm | 1.000 | 1.000 | (200) | 200 | (200) |
| Arsenic | | | | | | |
| inorganic | ppm | 50 | 50 | 50 | 50 | (50) |
| organic | ppm | 100 | 100 | 100 | 100 | (100) |
| Bromine | ppm | 200 | (200) | 200 | 2.500 | (200) |
| Calcium | % | 2 | 2 | 1 | 1.2-4.0 | 2 |
| Copper | ppm | 100 | 25 | 250 | 300 | 800 |
| Fluorine (mature animals) | ppm | 40-50 | 60-150 | 150 | 200 | (40) |
| Jodine | ppm | 50 | 50 | 400 | 300 | 5 |
| Iron | ppm | 1.000 | 500 | 3.000 | 1.000 | (500) |
| Lead | ppm | 30 | 30 | 30 | 30 | (30) |
| Magnesium | % | 1.000 | 1.000 | 400 | 2.000 | (400) |
| Mercury | ppm | 2 | 2 | 2 | 2 | (2) |
| Phosphorus | % | 1 | 0.6 | 1.5 | 0.8-1.2 | 1 |
| Potassium | % | 3 | 3 | (2) | (2) | (3) |
| Selenium | ppm | (2) | (2) | 2 | 2 | (3) |
| Sodium Chloride | % | 4-9 | 9 | 8 | 2 | (3) |
| Zinc | ppm | 500 | 300 | 1.000 | 1.000 | (500) |

The maximum tolerable level of specific nutrients in a feed depends mainly on the species of animals. A horse tolerates e.g. copper more than 30 times better than a sheep. Water soluble vitamins, fatty acids and most amino acids are well tolerated by most animals.

The non specific needs of an animal are defined by the energy needs. The energy needs can be covered by carbohydrates, lipids as well as by protein. These 3 nutrients can be exchanged largely each by an other. In most feeds the carbohydrates are the main energy source. Energy from protein is often expensive and the use of lipids is despite the high energy concentration limited for technical reasons (handling, oxidation etc.).

An energy supply only slightly below the needs causes already a decline of production. Overfeeding with energy above the needs has as consequence a undesired deposition of lipids in the body tissues.

## 3. UTILIZATION OF THE NUTRIENTS

The utilization of the nutrients determines directly the necessary nutrient supply to meet the animals needs. A high utilization allows a smaller feed intake, if the needs remain constant. The utilization of the nutrients is usually devided into the digestibility and the utilization in the tissues (intermediate utilization). The utilization of the nutrients in the digestive tract must be considered separately for the ruminants and the monogastric animals like pigs or poultry. The reason of the difference between the two groups of animals is the existence of an active microflora in the rumen. In figure 2 the two digestive tracts and the absorbed nutrients are shown in a simplified diagram.

Figure 2: Digestion and absorption of the nutrients in ruminants and the monogastric animal

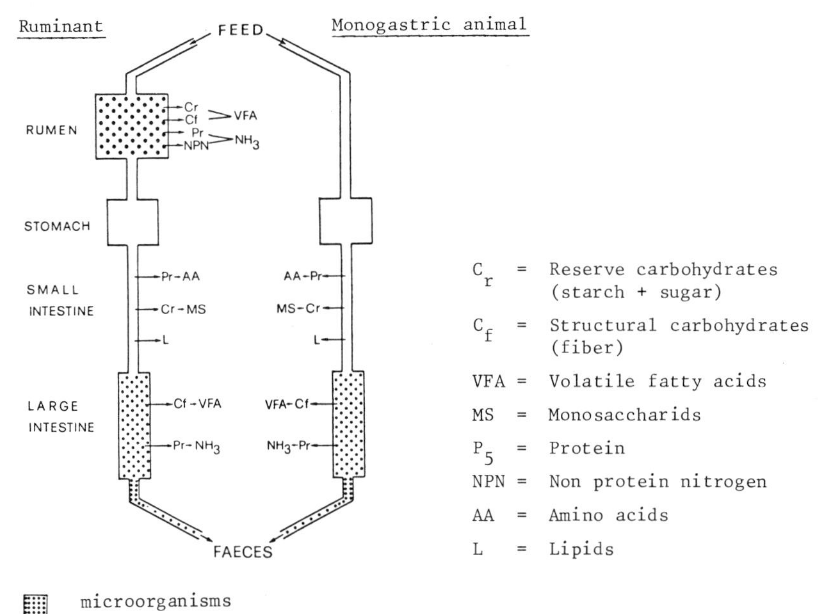

In the rumen reserve and structural carbohydrates are degraded by microbial fermentation into monosaccharids and then into volatile fatty acids (mainly acetic and propionic acid). Protein is partly degrated into aminoacids and into $NH_3$. But there is also growth of microorganisms in the rumen and therefore also a formation of microbial protein, vitamines etc. The microbial fermentation in the rumen has the following main effects on the utilization of the nutrients compared with monogastric animals:

positive

- Degradation of cellulose and other cell wall carbohydrates (fiber).
- Formation of microbial protein of a high quality from protein of a low quality and non protein nitrogen.
- Formation of microbial vitamins.

negative

- Degradation of reserve carbohydrates into volatile fatty acids.
- Degradation of amino acids into $NH_3$.
- Hydration of fatty acids and formation of branched fatty acids.

The further processes in the digestive tract of ruminants and monogastric animals are rather similar. The degradation of reserve carbohydrates, protein and lipids in the stomach and in the intestines is the result of the endogenous digestive secretions. The absorption of most nutrients takes place in the small intestine. In the large intestine the absorption of water, volatile fatty acids and some vitamines is of importance.

The further utilization of the absorbed nutrients (protein, carbohydrates and lipids) in the tissues is largely determined by the biochemical reactions and the type of production of the animal. In table 3 a crude attempt has been made to quantify the utilization of the nutrients in the digestive tract and in the intermediate metabolism.

Table 3: Utilization of the nutrients

| nutrient | utilization in the digestive tract | | utilization in the tissues |
|---|---|---|---|
| | monogastric animals | ruminants | |
| N-containing substance | | | |
| protein-high quality | *** | * - *** | *** |
| -low quality | * | * - *** | *(**) |
| non-protein-N (nucleic acids included) | 0 - * | * - *** | 0- [only ruminants - **] |
| carbohydrates | | | |
| reserve carbohydrates | *** | ** | *** |
| cell wall constituents (fiber) | * | ** | *** |
| lipids | | | |
| fatty acids C 20 or unsaturated | *** | ** | *** |
| fatty acids C 20 (saturated) | * - ** | * - ** | *** |

*** high; ** medium; * low; 0 no

In general the digestibility of protein of high quality, the reserve carbohydrates and the lipids is better in monogastric animals than in ruminants. Non protein-N (nucleic acids) can only be utilized by the microorganisms in the rumen and the utilization of cell wall constituents is better in ruminants

## 4. SINGLE CELL PROTEIN (SCP) IN ANIMAL NUTRITION

SCP can be utilized in animal nutrition, if it fullfills conditions, which are presented in table 4:

Table 4: Factors influencing the use of SCP in animal production

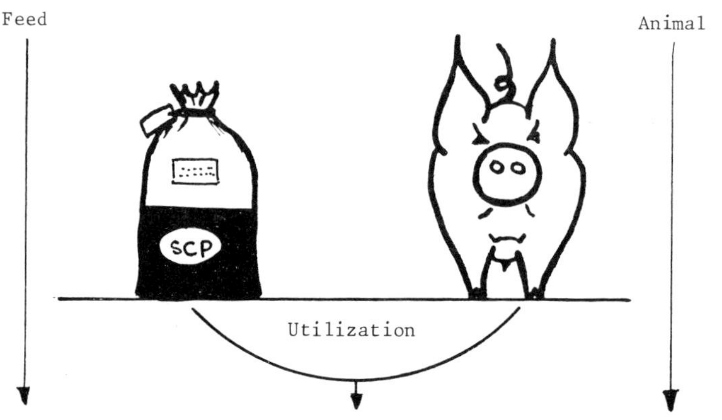


. High energy content

. High and equilibrated content of specific nutrients

. Low content of undesirable substances

. simple storage and good technological properities

. low price

. High digestibily

. High tissue-utilization

. no diarrhoea

. Good acceptance

. No specific effects

In competition with the other feedstuffs SCP must have a high content of specific nutrients (mainly essential amino acids) and a high energy content in relation to its price. The specific nutrients must furthermore be well equilibrated or complement to other feedstuffs. The use of SCP in the rations of domestic animals is possible in higher amounts, if it does not contain large amounts of undesirable or toxic substances. In this context the nucleic acids, non-protein-N and minerals from fermentation must be mentioned. Nucleic acids amount in SCP to about 15 % of N, but are only utilized by ruminants in higher amounts.Like all other feedstuffs SCF should have good technological properities (mixing, pelleting, storage etc.) The administration of SCP into the rations should not limit the voluntary feed intake of the animals and have no detrimental effect on the utilization of the nutrients of the ration.

# TOXICOLOGICAL EVALUATION OF SCP PRODUCED FROM WHEY

U. Schoch and Ch. Schlatter
Institute of Toxicology, Swiss Federal Institute of Technology and
University of Zurich, CH-8603 Schwerzenbach

## Summary

SCP produced from whey was evaluated for causing potentially toxic effects in animals. In a feeding experiment with rats no clear-cut effects could be detected. The measurements of the body weight, the organ weights, and some hematological and clinical chemistry values did not result in significant differences between the treated and the control group. In the mutagenicity test according to Ames a very slightly mutagenic activity of SCP crude extract was measured. But by comparison with other foods a negative effect to animals fed with SCP could be excluded. From the toxicological point of view the tested SCP powder from whey permeate seems to be a safe product.

## 1. Introduction

Feedstuffs containing single cell protein (SCP) differ in many ways from conventional feed. Besides protein they contain considerable amounts of nucleic acids, phosphates, lipids, and possibly biologically active, unmetabolized ingredients of the respective substrates. Additionally, the microorganisms used for fermentation can sometimes synthesize secondary metabolites with toxic properties. So the applicability of feedstuffs composed by SCP has firstly to be verified.

In this paper the results of the toxicological examination of SCP produced from whey are presented. In the first part the results from a feeding experiment with SCP-powder extracts are shown.

There were two questions to which the answers were hoped to be found by means of a subchronic feeding experiment with rats:

1) Are there still unknown substances metabolized by the yeasts to potentially toxic agents in the SCP-powder?
2) How hazardous are these substances and what are the consequences for the health state of the fattening animals?

In the second part the results about mutagenic activities in SCP from whey are discussed.

## 2. Experimental

- Calculation of daily SCP-intake: The factors used as a base for the calculation of the daily SCP-intake are listed up in figure 1.
- Single cell protein: The organism used for fermentation was the yeast Trichosporon cutaneum. The arguments for the use of that specific yeast and the technological data employed in the production of the examined spray-dried SCP-powder were reported by Prof. Puhan, Mr. Halter, and Mr. Käppeli at the same workshop.
- Extraction procedure: Because of the large daily dose an extraction of the powder was performed as described in figure 2. A procedure analogous to the classical mycotoxin analysis was used to get a defatted organic-soluble crude extract containing all potential toxins produced by the yeasts.

Figure 1: Daily SCP-intake during the feeding trial

| | |
|---|---|
| FATTENING ANIMALS: | PIG 100 KG BODY WEIGHT (B.W.) |
| DAILY FOOD-SUPPLY: | 1/40 OF B.W. ≙ 2,5 KG/100 KG B.W. |
| PART OF WHEY-SCP: | 10 % OF THE DAILY DIET ≙ 0,25 KG |
| FACTOR: | 100 |
| DAILY INTAKE: | 0,25 KG/KG B.W. X DAY |
| | ≙ 75 G SCP-POWDER/RAT (300 G) X DAY |

- Feeding experiment: These extracts were mixed with normal ground chow for laboratory rats. A daily portion of 75 g per rat of SCP powder as crude extract was offered to the animals. The animals used for the feeding experiment were Sprague-Dawley rats, kept under standarized conditions in groups of 10 young adult males. The following parameters have been determined: body weight, organ weights, hematological and clinical chemistry data.

- Mutagenicity: The mutagenic potential of SCP crude extract was evaluated by using the Salmonella typhimurium/liver microsome test developed by Ames et al. (1975).

Figure 2: Extraction procedure

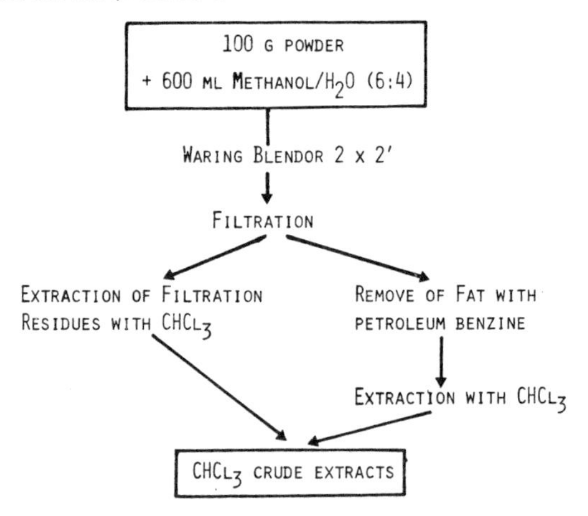


## 3. Results and discussion

A very sensitive parameter for monitoring toxic effects of test substances in the animals is the body weight. The increase in weight is shown in figure 3. Each point on the curves represents the average value of 10 male animals. The measurements at day 5 and 10 showed significant differences between the control and the treated group. The smaller increase in weight of the SCP-treated group can be attributed to the different composition of the rat chow at the beginning of the four week feeding period. However these differences did not exceed 10 % and are assumed to be of no biological relevance. After habituation to the SCP diet by the animals, the increase in body weight per time unit is the same in both groups. The difference seen in the first days could not be compensated for.

Figure 3: Body weight

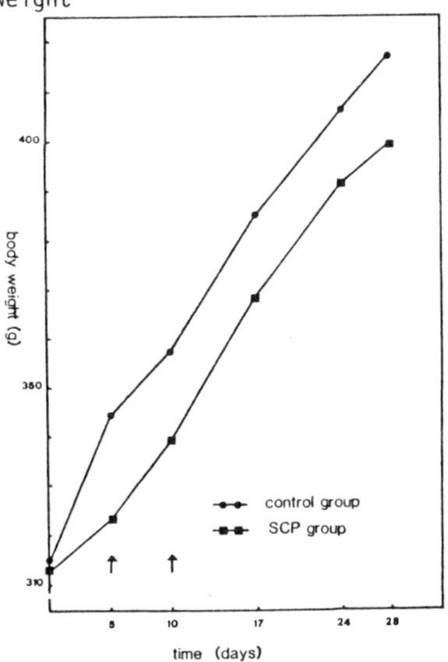


The mean values of the organ weights determined after ending of the feeding period are summarized in table 1.

Table 1: Relative organ weights

| | Liver | Kidneys | Spleen | Testes |
|---|---|---|---|---|
| Control Group | 3.85 ± 0.21* | 0.62 ± 0.02 | 0.19 ± 0.02 | 0.90 ± 0.08 |
| SCP Group | 4.18 ± 0.22 | 0.63 ± 0.03 | 0.19 ± 0.02 | 0.87 ± 0.08 |
| P | 0.001 | 0.297 | 0.455 | 0.193 |

*) $\bar{x} \pm s$

Macroscopic examination of the organs did not show any alterations. The liver weights of the control group were significantly lower than those of the SCP group. The statistically significant difference has to be considered to be the result of a slightly elevated functional liver activity. Histological examination of liver tissue of SCP treated rats did not show any pathological changes. In other organs such as kidneys, spleen, and testes, no differences in weight could be detected.

The hematological data determined in a Coulter Counter are shown in table 2. There is no indication of an alteration in any of the hematological parameters by feeding SCP crude extracts equivalent to 0,25 kg SCP powder per kg body weight and day.

Table 2: Hematological data

| | WBC ($\times 10^9$/L) | RBC ($\times 10^{12}$/L) | HGB (G/DL) | HCT (%) | MCV (FL) | MCH (PG) | MCHC (G/DL) |
|---|---|---|---|---|---|---|---|
| CONTROL GROUP | 9,7 ± 1,7 | 7,46 ± 0,3 | 16,6 ± 0,8 | 40,1 ± 1,5 | 55 ± 1 | 22,4 ± 0,7 | 41,5 ± 0,9 |
| SCP GROUP | 9,0 ± 1,0 | 7,49 ± 0,7 | 16,8 ± 0,9 | 41,3 ± 2,0 | 56 ± 3 | 22,6 ± 1,3 | 40,7 ± 1,3 |
| P | 0,137 | 0,440 | 0,340 | 0,073 | 0,067 | 0,337 | 0,071 |

The corpuscular constituents of the blood were separated by centrifugation and the plasma was used for the determination of clinical chemistry data. They are summarized in table 3.

Table 3: Clinical chemistry data

| | GOT (U/L) | GPT (U/L) | CHOL (MMOLE/L) | PROTEIN (G/L) | APHOS (U/L) | UREA (MMOLE/L) | LDH (U/L) | CK (U/L) | CREATININE (MMOLE/L) |
|---|---|---|---|---|---|---|---|---|---|
| CONTROL GROUP | 43 ± 3 | 41 ± 5 | 2,5 ± 0,2 | 88,1 ± 2,3 | 343 ± 56 | 6,4 ± 0,4 | 111 ± 38 | 64 ± 15 | 46,7 ± 3,3 |
| SCP GROUP | 44 ± 6 | 43 ± 4 | 2,2 ± 0,3 | 88,0 ± 2,3 | 305 ± 64 | 6,2 ± 0,7 | 109 ± 30 | 65 ± 16 | 54,2 ± 4,6 |
| P | 0,138 | 0,199 | 0,016 | 0,458 | 0,085 | 0,239 | 0,441 | 0,455 | 0,0003 |

CHOL: CHOLESTEROL

APHOS: ALKALINE PHOSPHATASES

The enzymes glutamic-oxalacetic-transaminase (GOT) and glutamic-pyruvic-transaminase (GPT) are good indicators of liver injury. The values of these two enzymes in the SCP treated group are normal. This fact is a further evidence for the small relevance of the slightly increased liver weights in the treated animals, because serious liver damage would impair the function of the organ.

The decreased cholesterol values were statistically significant but the differences to the control values were too small to be considered as pathological.

The kidneys regulate the plasma concentration of creatinine. In our experiment the values for creatinine in the SCP group were significantly increased. The situation can be characterized in the same way as the one with cholesterol just mentioned before. Even if the kidney is the only organ regulating plasma and urine creatinine concentration there are no clear signs for injuries to the kidneys because the values for urea and kidney-weight are normal.

The determinations of alkaline phosphatase, proteins, lactic dehydrogenase, and creatine kinase did not show any differences between the control and the treated group.

The principle of the mutagenicity test of Ames et al. (1975) is based on histidine auxotrophic (his-) Salmonella strains which are converted through the influence of a mutagen to prototrophic (his+) strains. That means: The Salmonella can only form colonies or revertants on a histidine free medium if they re-obtain the ability to synthesize histidine by mutation.

Certain substances are converted to mutagens by metabolic activation in mammalians. The enzymes, responsible for that transformation do not exist in bacteria. For this reason the activation systems have been isolated from rat liver homogenates by differential centrifugation. The supernatant fractions contain the enzymatically active liver microsomes and are called S-9. With this S-9 fraction it is possible to detect indirectly acting mutagens by combining the activating enzymes, the bacteria and the testsubstances.

The result of the Ames-test is considered to be positive if the number of revertants induced by the test substance is at least twice as high as the number of revertants of the background.

In our experiments we first tested increasing concentrations of crude extract from 0,01 - 10 mg per plate. All the strains (TA 98, 100, 1535, 1537, and 1538) were used with and without addition of S-9 fraction. The results showed only in the samples with strain TA 98 and TA 1537 slightly higher numbers of revertants indicating a light mutagenic activity in the SCP crude extract. All the other trials showed negative results.

In a second experiment using higher concentrations of crude extract with strain TA 98, a linear dose-effect-curve resulted. The mutation rate at the 20 mg level was about 4 (Figure 4).

Figure 4: Mutagenicity of SCP crude extract in the Ames test

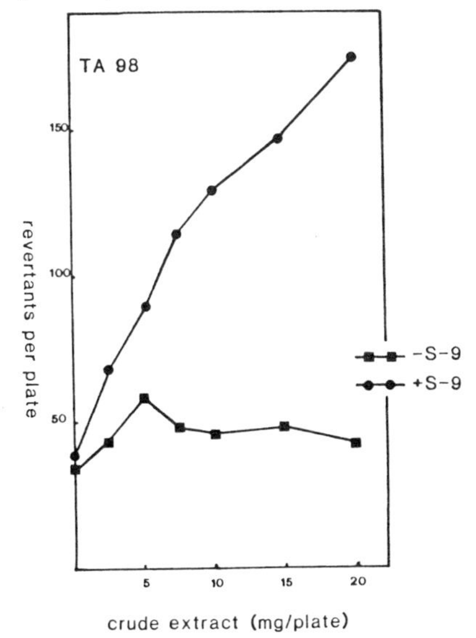

Based on these results it can be concluded that there is evidence for the existence of an indirectly acting mutagen in the crude extract.

What is the importance of this fact? In fattening animals no genetic effects are to be expected, especially when the short life expectancy is taken into consideration. In addition the mutagenicity is so weak that no tumorigenic response is expected, even if the compound would be fed until the whole life of a pig, that is approximately 15 years. Effects on human consuming the meat of such animals are difficult to estimate because of the unknown distribution of the substance in the animal body. Mutagenic activities in the Ames-test are known for other foodstuffs (Commoner et al., 1978; Spingarn and Weisburger, 1979; Wang et al., 1982) for instance grilled meat. Comparing the mutagenic activities of SCP with that of roasted foodstuffs, there results a lower mutagenic stress due to a daily direct consumption of 250 g SCP by man than caused by a meat consumption of 100 g per day.

## 4. Conclusions

Summarizing these results, it can be therefore concluded, that the feeding of SCP extract to rats did not provoke any effects indicative of the presence of toxins. In the Ames test a very slight mutagenic activity was measured, but by comparison with other foods a negative effect to the fattening animals or meat consumers could be excluded.

From the toxicological point of view SCP powder fermented from ultra-filtrated whey permeate by the yeast Trichosporon cutaneum seems to be a safe product.

## 5. Literature

Ames B. et al., Mut. Res. 31, 347 - 364 (1975).
Commoner B. et al., Science 201, 913 - 916 (1978).
Spingarn N.E. and Weisburger J.H., Cancer Letters 7, 259 - 264 (1979).
Wang Y.Y. et al., Cancer Letters 16, 179 - 189 (1982).

# MILK ULTRAFILTRATION PERMEATE FERMENTED BY YEAST: NUTRITIVE VALUE FOR GROWING PIGS

C. WENK
Institute for Animal Production, Nutrition Group, ETH Zentrum, CH-Zürich

## 1. INTRODUCTION

Whey is a common feedstuff for growing pigs. Its content of the main nutrients, calculated on a dry matter base, is similar to that of barley, but the high water content of whey limits the ingestion of whey nutrients in the diet. Modern technological methods (ultra filtration) allow, that most of the milk proteins can be used for cheese production. Consequently the by-products, called UF-permeates,contain less protein, causing a lower feeding value for the pig. The organic matter of UF-permeate contains mainly lactose which can cause diarrhoea in the growing pig and some fat.

PUHAN et al. (1983) performed a series of semi industrial experiments to study the fermentation of UF-permeate by yeast. Aim of the experiments was beside others to increase the feeding value of UF-permeate.

In feeding and digestibility experiments with growing pigs we tried to measure the nutritive value of these products.

## 2. COMPOSITION OF THE FERMENTED UF-PERMEATES (FP)

In 3 experiments with growing pigs we tested 3 different products: In table 1 the nutrient content of the fermented permeates (FP) is compared with unfermented permeate and whey.

Table 1: Nutrient content of whey and permeates

| | | whey | permeate | fermented permeate (FP) exp. 1 | exp. 2 (partly concentrated) | exp. 3 (dried) |
|---|---|---|---|---|---|---|
| Dry matter | g/kg | 48 | 46 | 35 | 91 | 44/967 |
| in dry matter | | | | | | |
| organic matter | g/kg | 903 | 893 | 659 | 710 | 840 |
| Crude protein | g/kg | 125 | 6 | 380 | 323 | 152 |
| Non-protein-N (% of crude-protein) | | - | - | 18 % | 15 % | 32 % |
| Soxhlett fat | g/kg | 8 | 16 | 65 | 20 | 66 |
| Lactose | g/kg | 770 | 870 | 14 | 30 | 248 |
| gross energy | MJ/kg | - | - | - | 16,7 | 16,1 |

- not analyzed

A direct comparison of the 3 FP is not possible. In the first two experiments a complete fermentation was strived. High crude protein and low lactose values are the consequence. Striking is the extremely high ash content, caused by fermentation additives. In experiment 3 FP was dried. Because the product was only partly fermented, crude protein content was lower and lactose content higher than in the previous experiments. The ash content could also be reduced by using organic acids during fermentation. The extremely high amount of non-protein-N could not be explained.

## 3. FEEDING AND DIGESTIBILITY TRIALS WITH GROWING PIGS

In a first experiment the voluntary feed intake of a FP sample was tested with 4 castrates of about 50 kg body weight. Feed intake of FP was compared with whey. As little as 4 litres of fermented permeate in addition to an ordinary compound feed caused diarrhoea in all animals, probably due to a very high phosphate content of FP. The control ration with 8 litres of whey caused no problems.

FP contains usually more than 95 % water. Therefore its application in practical pig feeding is limited. In experiment 2 we used a by hyper filtration partly concentrated sample of FP still containing some lactose.

In table 2 the experimental design and the results of the digestibility trials are given.

Table 2: Experiment 2: results

| Treatment | | 1 | 2 |
|---|---|---|---|
| basal ration | % 1) | 82 | 100 |
| fermented permeate | % 1) | 18 | - |
| animals | | 4 | 4 |
| body weight | kg | 75-89 | |
| digestibility of energy | | | |
| - whole ration | | 0.810 (0.003) | 0.826 (0.002) |
| - fermented permeate | | 0.74 | - |

1) dry matter base ( ) standard deviation

The animals ingested the FP unwillingly, probably because of the fast destruction during storage, but no disturbance could be observed. The calculated digestibility of energy of FP alone amounted to 0.74.

In experiment 3 a sample of dried FP, which was only partly fermented (see table 1), was mixed with a compound feed in high percentage in order to study the influence of FP and the nutritive value in digestibility trials with growing pigs.

Table 3: Experiment 3: results

| Treatment | | 1 | 2 | 3 |
|---|---|---|---|---|
| basal ration | % 1) | 100 | 70 | 50 |
| fermented permeate | % 1) | - | 30 | 50 |
| animals | | 6 | 6 | 6 |
| body weight | kg | | 24-48 | |
| digestibility of energy | | | | |
| - whole ration | | 0.782 (0.010) | 0.805 (0.015) | 0.809 (0.019) |
| - fermented permeate | | | 0.86 | 0.84 |

1) dry matter base ( ) standard deviation

Feed intake in treatment 3 was smaler than in the other treatments. But no diarrhoea could be observed. Digestibility of the energy of FP was far higher than in experiment 2, probably due to the better conservation of the dried compared to the concentrated material. It amounted to 0.84 in treatment 2 with 30 % FP in the ration. The further increase of FP to 50 % did not alter the digestibility positively.

## 4. CONCLUSIONS

Ultrafiltered permeate of milk contains a high amount of water (> 95 %). Organic matter consists mainly of lactose, which can cause diarrhoea in growing pigs. With a yeast fermentation of permeate the protein content can be increased, but the nutrient content is diluted.

In experiment 3 fermented permeate (FP) was only partly fermented. Digestibility of the energy was high and amounted to about 0.85. The nutrient content, calculated on a dry matter base, is similar to barley. On the base of fresh FP 1 kg barley would correspond to more than 20 litres FP and the cost of 1 liter FP should not exeed SFr. -.04,all production and transport costs included.

Due to the high water content and the difficulties with conservation fresh FP could not be fed to growing pigs in high amounts (experiments 1 and 2). It can be concluded that the nutrients of fresh FP are of a high quality and that the pig is able to utilize them well. Because of the high water content and the difficulties with conservation the use of FP will only be possible under favourable conditions (continuously fresh material, no transportcosts, no manure problems in the piggery).

# METHODS OF EVALUATION OF ENERGY AND PROTEIN VALUES FOR PIGS OF THREE YEASTS GROWN ON ALKANES

C. FEVRIER and D. BOURDON

I.N.R.A., Station de Recherches sur l'Elevage des Porcs
Saint-Gilles, F-35590 L'HERMITAGE

**Summary**

Digestible energy and protein values of three Candida alkane yeasts, C.lipolytica (B.P. France), C. tropicalis (B.P. Great Britain) and C. paraffinica (Dainippon, Rumania) were determined. They were introduced at three levels, 18, 36 and 54 p., in replacement of maïze starch in a protein-free semipurified diet. Digestible energy values, calculated by regression analysis, were : 4181, 4411 and 4417 kCal/kg of dry matter, respectively. Metabolisable energy contents for the first two yeasts were 3898 and 4008 kCal. Protein digestibility values were 90,6, 90,9 and 89,9 p. cent. Energy digestibility of the fattest yeast C.tropicalis 82,4 p.cent was lower than the others : 87,1 and 87,8 p.cent, in the order above.

## 1. INTRODUCTION

Protein value and possible toxicity of alkane yeasts have been the subject of a large number of studies, but few of them took into account the energy value, especially for the pig. Furthermore, the methods of determination, computation or direct measurement, were not always mentioned (3, 4, 6). Now, the energy value is the most important nutrient in the computerization of diets. We bring here our contribution to the determination of this value for three Candida yeasts. C. lipolytica (Cl) grown on gas oil, by B.P. France, in Lavera, C. tropicalis (Ct), grown on n-paraffins by B.P. Great Britain and C. paraffinica (Cp), grown on n-Paraffins by Dainippon (Japan) in Curtea de Argues in Rumania. Although these yeasts are nolonger on the market, the method used for the determination of their energy values is still useful for other SCP.

TABLE 1. -Gross chemical composition and main amino acids content of alkane yeasts used in the experiments (p.1000 g of dry matter).

| Yeasts | C.lipolityca | | | C. tropicalis | | | C. paraffinica | |
|---|---|---|---|---|---|---|---|---|
| Dry matter % | 94,05 | | | 94,68 | | | 95,12 | |
| Gross Energy | 4805 | | | 5351 | | | 5029 | |
| Crude Fat | 20 | | | 90 | | | 74 | |
| Ash | 79 | | | 67 | | | 75 | |
| Crude protein | 704 | | | 632 | | | 511 | |
| Lysine | 55 | 7,8 | g/16g N | 47 | 7,4 | g/16g N | 35 | 6,8 |
| Methionine | 11 | 1,6 | | 11 | 1,8 | | 7 | 1,4 |
| Cystine | 6 | 0,9 | | 7 | 1,1 | | 5 | 1,0 |
| Tryptophane | 9 | 1,3 | | 9 | 1,4 | | 7 | 1,3 |
| Threonine | 38 | 5,4 | | 31 | 4,9 | | 27 | 5,5 |
| References | (4) | | | (4) | | | INRA, 1983. | |

## 2.MATERIAL AND METHODS

In a protein free semi purified diet (PF) composed of maïze starch (57,56), maïze oil (2), wood cellulose, Colmacel F2 (6), mineral (2,44) and vitamin premix (2), the yeasts, of which the compositions are sumarized in table 1 were introduced at the 18, 36 and 54 per cent levels, replacing the starch. Possibly, the diets were fortified with limiting aminoacids (NRC, 1973). Each diet was given to 4 Large-white pigs, kept in metabolism cages for a collecting and measuring period of 10 days, preceded by an adaptation period of 7 days for the equalization of the intake levels. Diets were given wet in three meal a day. The use of three levels of yeast offers the possibility to check the lack of interaction with PF diet. Energy values are then calculated by regression according to the equation :
$Y/X_2 = a\ X_1/X_2 + b$ from 12 measurements for every yeast : with Y, a and b ; digestible energy of total diet, yeast and PF diet, respectively, in kCal/kg ; $X_1$ and $X_2$ : percentage of yeast and PF. Cl and Ct were tested in the C.N.R.Z., in Jouy en Josas, and Cp with SPF pigs, in the new station in Saint-Gilles.

TABLE 2. - Mean values of daily performance, apparent energy (ED) and crude protein digestibility (CPD) and nitrogen retention by pigs given several alkane yeast diets, and their digestible energy (DE)

| | | Experiment 1 | | | | | | Experiment 2 | |
|---|---|---|---|---|---|---|---|---|---|
| Initial weight | | 31,8 | | | | | | 41,8 | |
| Period term | | 10 | | | | | | 10 | |
| Yeast | | C. lipolytica | | | C. tropicalis | | | C. paraffinica | |
| Level of yeast | % | 18 | 36 | 54 | 18 | 36 | 54 | 18 | 36 | 54 |
| **Chemical analysis** | | | | | | | | | | |
| Dry matter | % | 91.3 | 92.7 | 91.7 | 91.3 | 92.3 | 92.1 | 88.8 | 89.8 | 90.7 |
| In dry matter | | | | | | | | | | |
| .Ash | % | 3.8 | 5.7 | 6.6 | 3.6 | 4.8 | 5.7 | 5.2 | 6.3 | 7.5 |
| .Crude protein | % | 12.9 | 26.2 | 37.8 | 12.8 | 24.3 | 33.7 | 10.9 | 20.1 | 28.5 |
| .Gross Energy | Kcal. | 4272 | 4385 | 4561 | 4404 | 4561 | 4804 | 4206 | 4391 | 4565 |
| **Daily performance** | | | | | | | | | | |
| Dry M. intake | g | 1187 | 1205 | 1009 | 1279 | 1292 | 1013 | 1432 | 1441 | 1461 |
| Daily gain | g | 675 | 615 | 540 | 525 | 475 | 640 | 600 | 672 | 685 |
| Feed conversion (DM/weight gain) | | 1.77 | 1.98 | 1.88 | 2.47 | 2.82 | 1.59 | 2.42 | 2.15 | 2.14 |
| **Apparent digestibility** | | | | | | | | | | |
| Dry matter | | $91.0_a$ | $88.6_{ab}$ | $86.8_c$ | $91.2_a$ | $89.5_{ab}$ | $86.6_c$ | $91.1_a$ | $90.6_a$ | $88.5_b$ |
| . Energy | | | | | | | | | | |
| E.D. | % | 90.5 | 89.5 | 87.9 | 91.1 | 89.3 | 86.9 | 91.9 | 91.0 | 89.5 |
| DE Kcal/kg DM | | 3905 | 3925 | 4009 | 4011 | 4075 | 4173 | 3867 | 3998 | 4085 |
| $s_{\bar{x}}$ (1) | | ± 8 | ± 28 | ± 12 | ± 5 | ± 17 | ± 20 | ± 43 | ± 13 | ± 16 |
| CV | | (0,4) | (1,4) | (0,6) | (0,2) | (0,8) | (0,9) | (2,2) | (0,6) | (0,8) |
| . Protein | | | | | | | | | | |
| C P D | % | 90.0 | 91.5 | 90.4 | 89.6 | 91.7 | 91.3 | 88.8 | 91.3 | 89.3 |
| N R C (2) | % | 66.3 | 40.2 | 30.4 | 70.9 | 44.0 | 41.0 | 76.1 | 51.1 | 43.1 |
| N retained | g/j | 14.6 | 18.6 | 16.8 | 16.6 | 20.0 | 20.5 | 16.8 | 22.1 | 25.6 |

(1) $s_{\bar{x}}$ : Standard error of the mean ; CV coefficient of variation
(2) NRC : Nitrogen retention coefficient ; N retained/ N absorbed.

3. RESULTS AND DISCUSSION.

Pig performance and diet digestibility according to the level of yeast are given in the table 2. Although the performances went with a large standard deviation (30 g/d), it seemed they reflected the intake of digestible energy for Cl and Cp. On the other hand, the weak performance obtained with Ct for the lower level could be related with a less favorable balance in aminoacids, in spite of the addition of methionine (0,2%). The linear response of the digestible energy content to the level of yeast, with regression coefficient r = 0,999, allowed to calculate the proper values for yeasts. The results are summarized in table 3. For the Cl and Ct, the metabolisable energy values were calculated by the same way : 3898 and 4008 Kcal/kg, respectively. These values and those from the table 3 are in agreement with those proposed by Van der Wall and al (6) for Cl. But, in comparison to those given by Van Weerden (in 4), they were higher for Cl and lower for Ct. The higher level of lipids in Ct did not seem to be an advantage since the digestibility of energy was lower than the ones of Cl and Cp were. The explanation could be a low digestibility of long chain fatty acids. The determination of protein digestibility was easy since the yeasts were the lone source. The results obtained were in good agreement with those already published for different kinds of alkane yeasts (1,2,3,4,5,6).

TABLE 3. - Energy and protein values in dry matter of alkane yeasts.

| Yeast | | C. lipolytica | C. tropicalis | C. paraffinica |
|---|---|---|---|---|
| Gross energy | kcal/kg | 4805 | 5351 | 5029 |
| Apparent digestibility | % | 87.1 | 82.4 | 87.8 |
| Digestible energy | kcal/kg | 4181 | 4411 | 4417 |
| Crude protein | g/kg | 704 | 632 | 511 |
| Apparent digestibility | % | 90.6 | 90.9 | 89.8 |
| Digestible protein | g/kg | 638 | 574 | 459 |

In conclusion, the digestible and metabolisable energy values of the alkane yeasts are not always in relation with their fat content. They are also subordinate to the composition of these fats.

REFERENCES

1. FEVRIER, C. (1971). Essais sur l'évaluation de la valeur protéique des levures sulfitiques et d'alcanes (B.P.) dans l'alimentation des porcs. Journées Rech. Porcine en France. 3, 91-96.
2. HEINZ, T., SCHADEREIT, R., HENK, G.(1979). Untersuchungen zum Einsatz von Erdoldistillat-Futterhefe "Fermosin" in der Tierernährung 2. Verdaulichkeit der Nährstoffe von fermosin-Futterhefe bei Broiler (Huhn) und Mastschwein. Arch. Tierernahrung 29.2 - 81-91.
3. SCHULZ, E., OSLAGE, H., J., (1977). Microorganisms as protein feed in animal nutrition. Animal Research and Development. 6; 7-35.
4. SHACKLADY, C., A., GATUMEL, E. (1971). Safety in use and nutritional value of yeasts grown on alkanes. 2nd. World Congress of Animal Feeding IV.general reports. Madrid.
5. TCHACHEV, I., GRIGOROV, V., ONUFRIEV, P., RYZHKOV, A., (1976). Fodder yeasts from petroleum distillates in feeds for pigs. Svinovodstvo,Moscow, N°2 13-14.
6. VAN DER WALL, P., VAN HELLEMOND, K.K., SCHAKLADY, C. A., VAN WEERDEN,E.J. (1971). Yeast grown on gas oil in animal nutrition. 2. in rations for pigs. 10th Int. Cong.Anim.Prod., Versailles, Theme 1/Pigs.

# ECONOMIC CONSIDERATIONS REGARDING SCP IN ANIMAL FEEDING

S. THOMKE
Dept. of Animal Husbandry, Swed. Univ. Agric. Sciences,
Funbo-Lövsta Expt. Station, S-755 90 Uppsala

Abstract

Economic values of Pruteen, Toprina, C. utilis and Pekilo for pigs and of the white rot fungus (S. pulverulentum) for ruminants have been calculated by substituting these products by soybean oil meal (SBOM), fish meal and cereals according to their nutritive values. Toprina were calculated to be 130 % of SBOM, those of C. utilis and Pekilo to be 96 and 91 %, respectively, and that of white rot fungus to be 71 %.

## 1. INTRODUCTION

In animal production the choice of feedstuffs is primarily based on their nutritive value (i.e. energy value and protein content, amino acid profile) and costs. Other characteristics of the feedstuffs to be considered are palatability, handling, technological, physiological and balancing properties, effects on product quality, hygienic quality and toxicity. In addition to toxins, a number of substances are present in SCP at elevated levels in comparison with conventional feedstuffs, and some of them (eg. poly-β-hydvoxy-butyric acid) are known to decrease performance in monogastric animal species when administered in substantial amounts (Krogdahl, 1979). This contribution is aimed at demonstrating some principles in calculating the economic value of SCP-products and of SCP-treated ligno-cellulosic material.

## 2. MATERIAL

Four SCP-products with varying nutritional properties have been selected and information given by different authors on chemical composition and nutritive values for pigs has been collected in Table 1. As can be noticed the crude protein (CP) content is highest in Pruteen, medium in Toprina and lowest in Candida utilis and Pekilo yeast. The average content of metabolizable energy (ME) used here for the four SCP-products was 16.6; 17.0; 13.1 and 14.0 MJ per kg DM respectively. As is known the chemical composition of SCP varies with cultivation conditions, therefore these values differ from those of other investigations, which certainly may influence the final result. The principle for the economic calculations was to carry out substitutions, i.e. to calculate the amounts of traditional feedstuffs (soybean oil meal (SBOM) fish meal (FM) and cereals) which from a nutritional point of view correspond to a certain amount of the SCP-product. Price calculations are based on the quantities of traditional feedstuffs corresponding to a certain amount of SCP. The calculations were based on the energy value and the content of lysine (lys expressed as digestible lys, assuming the same digestibility as for CP). Lysine can be regarded as a suitable indicator amino acid (AA), since it often is the first limiting AA in cereal based pig rations and since a stepwise supply of lys has been proven to improve performance in SCP-based pig rations (Hansen, 1982). The content of other constituents such as vitamins and minerals compared with the traditional feedstuffs has not been included in the calculations, since their economic value is rather limited and not fully known for the different products.

Table 1. Chemical constituents and nutritive value of some SCP-products and traditional feedstuffs for pigs.

| | Crude protein | | Lysine | | Ash | ME, | Dig. | |
|---|---|---|---|---|---|---|---|---|
| | % of DM | Dig., % | g/16 g N | g dig. /kg DM | % of DM | MJ/kg DM | lys./ MJ | Ref.[a] |
| Pruteen | 80 | 88 | 5.8 | 40.9 | 1 | 17.1 | | B |
| | 82 | 84 | 5.9 | 40.8 | 0.5 | 16.7 | 2.5 | S |
| | 78 | 87 | 6.4 | 43.5 | 0.8 | 15.9 | | H |
| Toprina | 61 | 87 | 7.2 | 38.4 | 6.5 | 15.3 | | B |
| | 64 | 93 | 8.2 | 49.0 | 6.6 | 19.1 | 2.4 | S |
| | 58 | 90 | 7.1 | 37.1 | 6.9 | 16.7 | | H |
| C. utilis | 49 | 70 | 8.4 | 28.8 | 9.0 | 13.1 | 2.2 | B |
| | 55 | | | | 7.5 | | | S |
| Pekilo | 56 | 75 | 6.4 | 27.0 | 5.6 | 14.6 | 1.7 | B |
| | 48 | 68 | 6.1 | 19.9 | 6.4 | 13.3 | | H |
| S. pulverulentum[b] | 39 | 82 | 4.2 | 13.4 | 5.6 | 10.1 | | T |
| Maize | 10 | 80 | 2.7 | 2.2 | 1.5 | 15.6 | 0.14 | E |
| Barley | 12.5 | 85 | 3.8 | 4.0 | 2.8 | 14.2 | 0.28 | E |
| Soyb. oil meal | 51 | 90 | 6.2 | 28.5 | 7 | 14.7 | 1.44 | E |
| Fishmeal | 75 | 93 | 8.0 | 55.8 | 15 | 19.0 | 2.94 | E |

[a] B Breirem (1976); S Schulz & Oslage (1975); H Hansen (1981); E Eriksson et al. (1976); T Thomke et al. (1980)

[b] Nutritive values for sheep

Table 2. Amounts calculated to be nutritionally equivalent to the sums of protein source plus cereal given.

| | One kg DM of the respective SCP-products corresponds to the sum of: | | | |
|---|---|---|---|---|
| | Pruteen | Toprina | C. utilis | Pekilo |
| Fishmeal DM, kg | 0.74 | 0.73 | 0.51 | 0.41 |
| + maize DM, kg | 0.16 | 0.20 | 0.22 | 0.39 |
| Fishmeal DM, kg | 0.73 | 0.73 | 0.50 | 0.39 |
| + barley DM, kg | 0.19 | 0.23 | 0.26 | 0.47 |
| SBOM DM, kg | - | - | - | 0.81 |
| + maize | | | | 0.12 |
| SBOM DM, kg | - | - | - | 0.80 |
| + barley DM, kg | | | | 0.16 |
| | One kg DM of SBOM corresponds to the sum of: | | | |
| Resp. SCP DM, kg | 0.67 | 0.68 | 0.98 | - |
| + maize DM, kg | 0.23 | 0.21 | 0.12 | |
| Resp. SCP DM, kg | 0.66 | 0.66 | 0.97 | - |
| + barley DM, kg | 0.27 | 0.23 | 0.14 | |

As a representative of the group of upgraded ligno-cellulosic materials S. pulverulentum is used to calculate its economic value. This type of product is of interest primarily for ruminants. Its chemical composition and nutritive value is also given in Table 1. The calculation is therefore based on energy and digestible protein.

## 3. RESULTS

The results from the replacement calculations are collected in Table 2. On a dry matter (DM) basis one kg of Pruteen corresponds to 0.74 kg fishmeal + 0.16 kg maize, whereas corresponding quantities for one kg Pekilo are 0.41 and 0.39 kg DM of FM and maize, respectively. Using barley as the cereal part decreases the amount of FM very little. This is an effect of a somewhat superior AA profile of barley compared with maize. Since SBOM contains less dig. lys per MJ than three of the SCP-products, the replacement equation has to be changed, i.e. Table 2 gives the amounts of the respective SCP-product + cereal which corresponds to one kg of SBOM. The economic values of the SCP-products have been calculated (in CIF Rotterdam-prices for 1981/82) in US $ per 1000 kg (Table 3). The results show similar calculated values for Pruteen as for Toprina. This is because a higher CP content of the former and a superior AA profile is balanced by a somewhat higher energy value of Toprina. The economic values calculated have been put in relation to the prices of FM and SBOM. These relative values show that Pruteen and Toprina have a relative value of approximately 80 % of the FM price and 130 % of the SBOM price. There is a great difference between the calculated values of the SCP-products based either on FM or SBOM, due to the fact that the nutrients (lys and energy) are cheaper in FM than in SBOM according to the prices used in these calculations.

Finally, calculations on the economic value of the white rot fungus S. pulverulentum have been carried out by using SBOM and barley. One kg DM of S. pulverulentum corresponds to 0.68 kg SBOM + 0.02 barley, equivalent to 226 US $ per 1000 kg 90 % DM, which meens a relative value compared with SBOM of 71 %.

Table 3. Calculated economic values for SCP-products at 90 % DM, price in US $ per 1000 kg[a].

| Basis | Pruteen | Rel. | Toprina | Rel. | C. utilis | Rel. | Pekilo | Rel. |
|---|---|---|---|---|---|---|---|---|
| Fishmeal + maize | 360 | 80 | 362 | 82 | 269 | 61 | 255 | 59 |
| " + barley | 365 | | 373 | | 279 | | 274 | |
| SBOM + maize | 421 | 130 | 421 | 132 | 309 | 96 | 287 | 91 |
| " + barley | 407 | | 420 | | 305 | | 294 | |

[a] Average pirces CIF, Rotterdam, 1981/82 US $ per 1000 kg, Maize 182; Barley 205; SBOM 319; Fishmeal 450.

## 4. REFERENCES

Breirem, K. 1976. Encellprotein i foringa av slaktesvin og kyllinger. Inst. husdyrern. foringslära, Norges landbrukshögskole, Melding 179, Oslo, 65 pp.

Eriksson, S., Sanne, S. & Thomke, S. 1976. Fodermedeltabeller och utfodringsrekommendationer. LT:s förlag, Stockholm, 62 pp.

Hansen, J.T. 1981. Bioproteins in the feeding of growing-finishing pigs in Norway. Tierphysiol., Tierernährg. u. Futtermittelkde. 46, 182-196.

Hansen, J.T. 1982. Ibid, 47, 21-34; 35-42; 43-52.

Krogdahl, Å. 1979. Organiske forbindelser i encellprotein, effekter i enmagede dyr. Inst. fjörfe og pelsdyr, Norges landbrukshögskole, Melding 62, Oslo, 9 pp.

Schulz, E. & Oslage, H.-J. 1975. Analytische und tierexperimentelle Untersuchungen zur ernährungsphysiologischen Qualität von biotechnischem Protein sowie dessenErgänzungsmöglichkeiten. Sdrh. Ber. Landw. 192, 607-621.

Thomke, S., Rundgren, M. & Eriksson, S. 1980. Nutritional evaluation of the white rot fungus Sporotrichum pulverulentum as a feedstuff to rats, pigs and sheep. Biotechn. Bioeng. 22, 2285-2303.

# COMMUNITY GUIDELINES FOR THE ASSESSMENT OF NON-TRADITIONAL PRODUCTS OBTAINED THROUGH THE CULTURE OF MICROORGANISMS AND USED IN ANIMAL NUTRITION

S. MALETTO
Chairman of the Scientific Committee for Animal Nutrition,
Commission of the European Communities, Brussels
Facoltà di Medicina Veterinaria
Università degli Studi
via Nizza, 52
I - 10126 TORINO

The Guidelines were drawn up by the Scientific Committee for Animal Nutrition at the request of the Commission of the European Communities. Over a period of two years the Committee worked to establish a model which could be used for experiments, data collection, data processing, writing of reports and dossiers and interpretation of results.

The draft is based on the harmonizing of documents already prepared and discussed by committees set up by the EEC Member States and by national and international organizations, which had dealt with the matter in the past and kept abreast of the most recent scientific advances as regards safeguarding human and animal health and protecting the environment. In addition to giving references to the relevant Council Directives, the introduction to the Guidelines sets out the working philosophy involved in the preparation of essential documentation, upon which acceptance of these dossiers for evaluation purposes depends. It would appear that, in spite of their stringency, the guidelines are to be regarded as an "open system" for general application. Although some of the questions listed may be left unanswered when they are clearly superfluous, reasons must be given. The text also indicates that new research methods, derived from advances in scientific knowledge, may be introduced.

This "openness" means that the applicant will have to provide a dossier prepared on the basis of the most advanced research methods, and that the experts will have to base their evaluation on the most up-to-date criteria.

The concept of an "open system" also provides the scientific organizations with the opportunity of requesting further research to be carried out in respect of products already authorized, in order to take account of scientific and technical advances relevant to the safeguarding of human and animal health and the protection of the environment.

The guidelines are divided into four sections:

The first contains general criteria for the characterization of the microorganism, the culture medium, the processes involved in manufacturing, purification and checking of the purity and consistency of the product. It also deals with its physical and physico-chemical properties, its composition, commercial preparations and checking methods.

The second section lists the studies required to assess the product's nutritional properties, to be carried out on either laboratory, animals or target species. The purpose of this research is to establish the therapeutic indices for user animals and to discover what, if any, changes need to be made as regards the bromatological and technological qualities of the feedingstuffs produced.

The third section deals with studies to assess how safe a product is for the target species, and to evaluate the direct and indirect risks involved in using the product, as regards humans and the environment. The toxicological studies required for this purpose depend upon the type of product concerned, the animal species involved and the metabolism of the laboratory animal. This is the most critical section, given the difficulties involved in applying traditional toxicological models to the assessment of substances specifically intended for use as feedingstuffs.

The fourth section, entitled "Other Relevant Studies", refers to studies which (depending on the nature of the product) are intended to assess and prevent potential risks deriving from handling the product, e.g. allergic effects, irritation of the skin, the mucous membranes of the eye and the respiratory and digestive tracts.

# ROUND TABLE DISCUSSION

## EVALUATION AND RECOMMENDATIONS

### 1. Evaluation of the results of the COST-Workshop 83/84 held in Zurich, April 13-15, 1983.

The COST 83/84 programme is scientifically rather heterogeneous. It embraces two very different residuals from agriculture as feedstock for SCP-production (lignocellulosics / whey) and the testing of toxicology / nutritive value / acceptability of the products.

#### 1.1. Proteinization of lignocellulosic materials

Progress has been made in some directions (pre-treatment, strain evaluation and improvement, enzyme production and process development).

However, development of inexpensive, efficient processes, including a simple process for application on farms, has not yet been possible. The main reason for this failure is the fact that new approaches not only solve problems, but also demand that newly arising gaps be closed.

#### 1.2. Proteinization of whey

Unlike straw and other plant residues, whey represents an easily accessible feedstock for microbial growth. Feasibility of a yeast process has been established. However, the economy depends on many circumstances related to the structure of the dairy industry and economical conditions given in each particular case. Chances for an economical process are therefore only given if its layout meets the requirements of the given conditions. Nevertheless, no full size plant case study with economical considerations has been presented. However, in such a study, cooperation with the International Dairy Federation is desired.

#### 1.3. Nutritive value and toxicology

In the centre of nutrition studies are feeding trials in the target animal. However, toxicological studies with laboratory animals are not needed to the same extent as are needed in studies of human foodstuffs. Official guidelines and requirements should reflect this principal difference.

Due to the intensive interactions between microorganisms and substrates, tests for feeding value and toxicological properties should be performed on the final product.

Proteinized whey is considered as suitable for animal feeding; its nutritional value has been established, and the absence of toxic properties has been shown. Feeding studies and toxicological evaluation of the various lignocellulosic-SCP products are much less advanced. This is due to the incompleteness of process development, the poor consistency of the described product and the limited availability of the products. Moreover, PAG guidelines are more explicite and applicable than EC-guidelines.

## 2. Recommendations

### 2.1. Single Aspects

- Restrict the microbiological or chemical conversion of waste substrates to those that are readily available and/or not adequately used. Adhere to the proposals outlined for COST 84 bis.

- Consider the production of feed which is protein-enriched and at the same time improved with respect to digestibility.

- Concentrate on direct conversion into end products; avoid the microbial production of intermediates of a relatively low value.

- Restrict a high technology input to high value and products, but use defined waste materials in the technologies whenever appropriate.

- Concentrate on an overall use of all chemical compounds involved; avoid the creation of new waste problems.

- Use the specific facilities of a "concerted action" to conduct comparative studies if possible and have the state-of-the-art evaluated in a professional and critical way.

- Take the quality requirements defined for animal feed into an early consideration.

- Take the potential for a technical scaling-up into early consideration and develop new laboratory systems only if accompanied by convincing arguments.

- Take economical feasibility of novel developments into consideration, but do not allow these considerations to prevent testing of new hypotheses and principles.

### 2.2 Sectorial Aspects

* Proteinization of lignocellulosic material

Promotion of basic research on pretreatment, strain development and enzyme formation. Achievement of process development is only possible on the basis of a sound scientific background.

* Proteinization of whey

In certain cases, the large-scale production of SCP or proteinization as alternatives for the utilization of whey is the only solution to overcome a critical situation for the disposal of whey. Then a case study should be underetaken to reveal the economical costs of a potential, full-scale plant, compared with treating whey as waste water.

* Nutritive value and toxicology

Determination of accurate nutritive value and limited toxicological studies (e.g. mutagenicity tests) should be maintained in suitable forms for quality control of products of full scale plants.

# LIST OF PARTICIPANTS

ADLER, I.
Institut für Biotechnologie
ETH-Hönggerberg/HPT
Ch - 8093 ZUERICH
Tel: (01) 377 2195

ALLERMANN, K.
Institute of Plant Physiology
University of Copenhagen
Ø. Farimagsgade 2A
DK - 1353 COPENHAGEN K
Tel.: (01) 135915

ANDER, P.
Swedish Forest Products
Research Laboratory
P.O. Box 5604
S - 11486 STOCKHOLM
Tel.: (08) 224340

ARNOUX, Ph.
Générale Sucrière
Eppeville 80
F - 80400 HAM

BAN, S.
Faculty of Food Technology
and Biotechnology
Pierottijeva 6
Yugoslavia - 41000 ZAGREB

BARFOED, S.
D.D.S. - Biochemical Laboratory
Langebrogade 1
DK - 1001 COPENHAGEN
Tel: (01) 954100

BAYEN, M.
Prospective Department
Ministry of Industry and Research
1, rue Descartes
F - 75231 PARIS, Cedex 05
Tel: (01) 634.31.63 - 634.32.02
Telex: MIR 204 643 f

BERTRAND, D.,
I.N.R.A.
Laboratoire de Technologie des
Aliments des Animaux
Route de la Geraudière
F - 44072 NANTES, Cedex
Tel.: (40) 76.23.64
Telex: INRANTE 710.074

BLACHERE, H.
I.N.R.A.
17, rue de Sully
F - 21034 DIJON
Tel: (80) 65.30.12

BLUZAT, R.
Centre de Prospective et d'Evaluation
du Ministère de l'Industrie et de la
Recherche
1, rue Descartes
F - 75 231 PARIS, Cedex 05
Tel: (1) 634.33.33

BUNGAY, H.
Rensselaer Polytechnic Institute
204, Ricketts Bldg.
USA - TROY, N.Y. 12181
Tel: 518.270.6376

COTON, S.G.
Dairy Crest Milk Marketing Board
Thames Ditton
UK - SURREY KP7 OEL

DE LA TORRE, M.
Department of Biotechnology
Centro de Investigacion y de
Estudios Avanzados
I.P.N.
Apdo Postal 14-740
Mexico - 07000 MEXICO CITY

DOMSCH, K.H.
F.A.L. (Forschungsanstalt für Landwirtschaft)
Bundesallee 50
D - 3300 BRAUNSCHWEIG

DUPUY, P.
I.N.R.A.
17, rue de Sully
BV 1540
F - 21034 DIJON
Tel: (80) 65 30 12

DURAND, A.
I.N.R.A.
Station de Génie Microbiologique
17, rue Sully
F - 21034 DIJON
Tel: (80) 65.30.12

EISEN, L.
Commission of the European Communities, D.G. Information Market and Innovation
P.O. Box 1907 - JMO - B4/070
L - 2920 LUXEMBOURG
Tel: 43011 x 3164
Telex : 3423/3446 comeur lu

FAEHNRICH, P.
Institut für Biotechnologie
Kernforschungsanalge Jülich
Postfach 1913
D - 5170 JUELICH
Tel: (02461) 61.55.85

FERRANTI, M.-P.
Commission of the European Communities, D.G. Science, Research and Development - COST
200, rue de la Loi (SDME 03/63)
B - 1049 BRUSSELS
Tel: 235 11 11
Telex: 21877 Comeu B

FERREREO, G.-L.
Commission of the European Communities, D.G. Science, Research and Development
200, rue de la Loi (SDME 02/15)
B - 1049 BRUSSELS
Tel: 235.11.11
Telex: 21877 Comeu B

FEVRIER, C.
I.N.R.A.
Station Recherches Porcines
St. Gilles
F - 35590 L'HERMITAGE
Tel: (99) 64.62.63

FIECHTER, A.
Institut für Biotechnologie
Eidg. Techn. Hochschule
ETH-Hönggerberg/HPT
CH - 8093 ZUERICH
Tel: (01) 377.20.83

GASCHE, U.P.
Cellulose Attisholz AG
CH - 4708 LUTERBACH

GAVIN, M.
Bühler Brothers Laboratories
CH - 9240 UZWIL
Tel: (073) 50 11 11
Telex 77541 gbu ch

GUERRA-SANTOS, L.
Institut für Biotechnologie
ETH Zürich
Dörflistr. 75
CH - 8050 ZUERICH

HARTLEY, R.D.
Grassland Research Institute
UK - HURLEY, MAIDENHEAD
Tel: 062.882.3631

JOHANIDES, V.
Faculty of Food Technology and Biotechnology
University of Zagreb
Pierottiseva 6
Yugoslavia - 41000 ZAGREB
Tel: 041.440.422

JONAS, D.
Ministry of Agriculture, Fisheries and Food
Great Westminster House
Horseferry Road
UK - LONDON
Tel: 01.216.7453

KAM-THONG, G.
TECHNIP
Tour Technip
F - 92090 PARIS LA DEFENSE 6,
Cedex 23
Tel: (1) 778.2392
Telex: TCNIP A 612839 F

LAFORCE, R.
Institut für Biotechnologie
ETH-Hönggerberg
CH - 8093 ZUERICH
Tel: (01) 377.24.29

LADETTO, G.
Istituto Scienze degli Allevamenti
Via Nizza 52
I - 10126 TORINO
Tel: (011) 651.656

LARIOS DEANDA, G.
I.N.R.A.
17, rue Sully
F - 21034 DIJON
Tel: (80) 65.30.12

LEISOLA, M.S.A.
Lignocellulose Group
Department of Biotechnology
Swiss Federal Inst. of Technology
ETH-Hönggerberg
CH - 8093 ZUERICH

LINKO, M.
VTT - Biotechnical Laboratory
Tietotie 2
SF - 02150 ESPOO 15
Tel: 358.0.4565150

MALETTO, S.
Dept. Produzioni Animali, Ispezione ed Igiene Veterinaria
Via Nizza 52
I - 10126 TORINO
Tel: (011) 65.16.56

MOEBUS, O.
Bundesanstalt für Milchforschung
Hermann-Weigmann-Str. 1-27
D - 23 KIEL 1
Tel: (0431) 609.438

MONTENECOURT, B.S.
Department of Biology and the Biotechnology Research Center
Lehigh University
USA - BETHLEHEM, P.A. 18015

MORESI, M.
Istituto di Chimica Applicata e Industriale
Faculty of Engineering
Via Eudossiana 18
I - 00184 ROMA
Tel: (04) 47.514.53

MOULIN, G.
Chaire de Génétique et Microbiologie - I.N.R.A. - E.N.S.A.
Place Vialla
F - 34060 MONTPELLIER, Cedex
Tel: (67) 63.02.02

NALIN, F.
Générale Sucrière
F - 14630 CAGNY
Tel: (31) 84.26.02

OLSEN, J.
Institute of Plant Physiology
University of Copenhagen
Ø. Farimagsgade 2A
DK - 1353 COPENHAGEN
Tel: (01) 13.59.15

PAQUOT, M.
Departement de Technologie
Faculté des Sciences Agronomiques
Avenue de la Faculté
B - 5800 GEMBLOUX
Tel: (081) 61.29.58 - ext. 2304

PERINGER, P.
E.P.F.L.
Laboratoire de Génie Biologique
RAC-Changins
CH - 1260 NYON
Tel: (022) 61.54.51

PRENDERGAST, P.
Department of Microbiology
University College
IRL - GALWAY
Tel: 24411
Telex: 28823 UCG EI

PUHAN, Z.
Institut für Lebensmittel-
wissenschaft
ETH Zentrum
Ch - 8092 ZUERICH
Tel: (01) 256.53.68

REY, J.-P.
EPF de Lausanne
Ecublens
CH - 1015 LAUSANNE

REXEN, F.
Carlsberg Research Laboratory
Department of Biotechnology
DK - COPENHAGEN

RIJKENS, B.A.
Inst. for Storage and Processing
of Agric. Prod. (IBVL)
P.O.B. 18
NL - 6700 AA WAGENINGEN
Tel: 08370 - 19043
Telex: 45371

ROGERS, P.L.
School of Biotechnology
University of NSW
P.O.B. 1
Australia - KENSINGTON NSW 2033
Tel: (02) 662-2668

ROULET, N.
Office Fédéral de l'Education et
de la Science
Wildhainweg 9
CH - 3003 BERNE

SARRA, C.
Centro CNR
Alimentazione animale
Via Nizza 52
I - 10126 TORINO
Tel: (011) 65.16.56

SCHAEFFER, O.
EPF Lausanne
Institut de Génie Chimique
CH - 1015 ECUBLENS

SCHOBINGER, U.
Eidg. Forschungsanstalt
für Obst-, Wein- und Gartenbau
CH - 8820 WAEDENSWIL
Tel: (01) 780.13.33

SCHOCH, U.
Institut für Toxikologie
ETH - Uni Zürich
Schorenstr. 16
CH - 8603 SCHWERZENBACH
Tel: (01) 825.10.10

SCHUEGERL, K.
Institut für Technische Chemie
Universtät Hannover
Callinstr. 3
D - 3000 HANNOVER 1

SENEZ, J.C.
C.N.R.S.
Laboratoire de Chimie Bactérienne
31, chemin Joseph Aiguier
F - 13009 MARSEILLE
Tel: (91) 71.16.96
Telex: F 430 225 CNRSMAR

SPREY, B.
Institut für Biotechnologie
Kernforschungsanlage Jülich
Postfach 1913
D - 5170 JUELICH

STALDER, H.
Maschinenfabrik Gebr. Bühler AG
Bahnhofstrasse
CH - 9240 UZWIL
Tel: (073) 50 22 75
Telex: 77541 gbu ch

STEWART, C.
Rowett Institute
Greenburn Road
UK - BUCKSBURN ABERDEEN AB2 9SB
Tel: (0224) 71.27.51 - ext. 128

STRINGER, D.
ICI - Agricultural Division
UK - BILLINGHAM
Tel: 0642-553601

TEILHARD de CHARDIN, O.
Station de Génie Microbiologique
17, rue Sully
F - 21000 DIJON
Tel: (80) 65.30.12

THOMKE, S.
Dept. Animal Husbandry
Swed. Univ. Agric. Sciences
S - 75007 UPPSALA
Tel: (018) 15.20.70

ULMER, D.
Institute for Biotechnology
ETH-Hönggerberg/HPT
CH - 8093 ZUERICH

VON STOCKER, U.
EPF de Lausanne
EPF - Ecublens
CH - 1015 LAUSANNE
Tel: (021) 47.31.85

WATTENHOFER, R.
Ecole Polytechnique Fédérale
Génie biologique
Changins
CH - 1260 NYON
Tel: (022) 61.54.51

WENK, C.
Institut für Tierproduktion
Gruppe Ernährung
ETH Zentrum
CH - 8092 ZUERICH
Tel: (01) 256 3355

ZADRAZIL, F.
Institut für Bodenbiologie
Bundesallee 50
D - 3300 BRAUNSCHWEIG
Tel: 05373.7133 or 0531.59.63.58

ZADRAZIL, P.
Institut für Tierernährung
U. Dejvického Rybnicku, 10
CSSR - PRAHA-DEJVICE

YAZICIGLU, T.
Marmara Research Institute
P.O. Box 21
Turkey - GEBZE

ZURBRIGGEN, D.
I.G.C. - E.P.F.L.
Ch - 1015 ECUBLENS
Tel: (021) 47.31.74

# INDEX OF AUTHORS